# LEAVING THE SAFE HARBOR

THE RISKS AND REWARDS OF RAISING A FAMILY ON A BOAT

TANYA HACKNEY

LEAVING THE SAFE HARBOR
The Risks and Rewards of Raising a Family on a Boat

Copyright ©2021 by Tanya Hackney
All rights reserved, in all countries and territories worldwide.

Published by Ingenium Books Publishing Inc.
Toronto, Ontario, Canada M6P 1Z2
https://ingeniumbooks.com

Copyright fuels innovation and creativity, encourages diverse voices, promotes free speech, and helps create a vibrant culture. Thank you for purchasing an authorized edition of this book and for complying with copyright laws by not reproducing, scanning, or distributing this book or any part of it, in print or electronic form, without the express prior permission of the publisher. Please respect the hard work of the author and do not participate in or encourage the piracy of copyrighted intellectual property.

ISBNs

eBook: 978-1-989059-76-0
Paperback: 978-1-989059-75-3
Audiobook: 978-1-989059-77-7

*Book cover design by Jessica Bell via Ingenium Books*
*Book formatting by Ingenium Books*
*Photo of s/v Take Two by Bruce Vanderbilt, taken at Montserrat, 2016*
*Author photography by Debra Ferragamo-Hayes*
*Iconography via Shutterstock*

## PRAISE FOR LEAVING THE SAFE HARBOR

Hackney is a strong writer, at home describing her band of sailors successfully pulling together in a sudden squall or the disorienting effects of snorkeling in deep water. She presents a cleareyed assessment of her family ("The truth is that we more often resemble the cast of Gilligan's Island than an Olympic sailing team") that makes them sympathetic and accessible. Hackney writes about her parenting decisions in detail but without giving the impression that others should follow in her footsteps, allowing the reader to join the crew vicariously without feeling inadequate and to absorb the book's lessons without being bludgeoned by them. The blend of reminiscence and analysis makes for a satisfying read, both entertaining and insightful.

An engaging, thoughtful look at life on a sailboat.

*—KIRKUS REVIEWS*

*Leaving the Safe Harbor* is the most honest account of the cruising life that I've read: the highs, the lows, and the amazing rewards that come with a more adventurous life. This book proves, once again, that cruising families are truly special.

Yes, I absolutely LOVED this book!

—CAROLYN SHEARLOCK, AUTHOR OF *THE BOAT GALLEY*

Being a lifelong cruiser, (captain of *S/V Whensday*) I see a lot of folks try to escape to our lifestyle. They just sell out, jump on a boat and are back on land after a season or two. Success stories are rare. Success stories with the added facet of raising children are even more rare. Tanya's book should be mandatory reading for all of them.

So, you want to go from the magazine pictures of permanent cruising to the real thing? This book should be your very first step. With forty years of cruising, my wife and I can tell you that Tanya has captured the absolute essence of our culture out here. Relationship challenges, finance issues, and seamanship are all captured in unadulterated light and penned with magical wit.

—MICHAEL A. BARBER, AUTHOR OF *THE WATER: LIFE AND ADVENTURE*

*Leaving the Safe Harbor* will either inspire you to take a leap into the scary unknown and move onto a small boat with your family—or else it will renew your sense of gratitude for your cozy, dry, and comparatively uncrowded life on land.

Either way, *Leaving the Safe Harbor* will have you in turns holding your breath, laughing out loud, and staring at the pages in awe. This book is a fascinating insight into a wildly different way to raise a family, rich with life lessons on courage, relationships and parenting that can benefit any of us.

I'm sure sailors, wanna-be sailors and parents looking for some escapism from the mundane will gobble it up.

—TORRE DEROCHE, AUTHOR OF *LOVE WITH A CHANCE OF DROWNING*

If you are forty-five or below and interested in living the dream with a family in tow, or you're older and just curious about what it is like to truly live and raise your children on a boat, this is your book.

What *Leaving the Safe Harbor* is not is a compendium of hair-raising sea stories, a travelogue, or an instruction manual on sailing. What it is—is a very personal and frank description of what decisions were made, why, and how they turned out with respect to the remarkable Hackney family over the years of their extensive travels.

Suffice it to say that the twists and turns of Hackney, her husband, and their children's lives together can truly be referenced as a work-in-progress.

A highly recommended read.

—REX COWAN, *FORMERLY OF S/V GENESIS*

Through the voice of Tanya, self-professed control freak and loving mother of five, *Leaving the Safe Harbor* propels the reader through the riveting, thought-provoking, and often humorous voyages her family undertakes. A sense of wonder and humor pepper the pages of this fantastically penned memoir. The reader will find themselves filled with emotions as they experience the wonders of the ocean and nature, enhanced by the profound love this family has for each other.

One is able to taste the ocean waves splashing their face, smell the sparkling sea, and feel the shining sun on their skin with her vivid recount of their adventures. I could really picture all the details she so eloquently weaves into their story. By the end, I found myself wanting to buy a catamaran and sail alongside this amazing family!

Overall, this is a superbly written memoir that I would rate a five out of five stars. This book is most appropriate for mothers who yearn for something more. Those who may need the inspiration to take the necessary steps to reach their own unique dreams. It is also appropriate for those people who like reading a memoir about love, family adventure, personal growth, and life at sea.

—DASHAINA GIBBS, NETGALLEY REVIEWER

Tanya's story is a perfect example of what can happen when "someday" becomes "what if," then "why don't we," and finally, "we're doing this." Should you read this book? I'll answer with a question—are you someone who often wonders if there's more to life than you're currently experiencing? If the answer is yes, then yes, you should.

—CELESTE ORR, AUTHOR OF
*TOGETHERNESS REDEFINED*

## CONTENTS

Crew List — xiii

Prologue: Staying Afloat — 1
*The Worst and the Best*

1. Rocking the Boat — 9
   *Big Dreams*
2. Uncharted Waters — 19
   *A Leap of Faith*
3. Sink or Swim — 31
   *Survival Skills*
4. Running a Tight Ship — 45
   *Discipline*
5. Learning the Ropes — 57
   *Making Mistakes*
6. Close Quarters — 69
   *Conflict Resolution*
7. Chock-A-Block — 87
   *Collecting Verbs*
8. All Hands on Deck — 103
   *Teamwork*
9. Batten Down the Hatches — 123
   *Hardship and Hope*
10. Getting Shipshape — 139
    *Organized Chaos*
11. See Which Way the Wind is Blowing — 153
    *Decision Making*
12. Plumbing the Depths — 163
    *Gratitude and Awe*
13. Ships Passing in the Night — 177
    *Friendships Afloat*
14. Troubled Waters — 191
    *Patience*

15. On the Right Tack — 205
    *Give and Take*
16. Smooth Sailing — 221
    *Simple Appreciation*
17. Course Corrections — 231
    *Flexibility*
18. Safe Harbor — 245
    *Letting Go*
    Epilogue: In the Offing — 261
    *New Dreams*

Glossary of Nautical Terms — 267
Acknowledgments — 279
About the Author — 281

*For Jay, our intrepid captain and for the crew of Take Two*

Twenty years from now, you will be more disappointed by the things you didn't do than by the ones you did do. So throw off the bowlines. Sail away from the safe harbor. Catch the trade winds in your sails. Explore. Dream. Discover.

MARK TWAIN

## CREW LIST
### WHO'S WHO ON TAKE TWO

**Jay, Captain and Chief Engineer.** The problem-solver, the magic genie who funds the dream, and the introverted computer genius with an adventurous side. He grew up sailing and served as crew on race boats. He has two full-time jobs, working as a consultant and keeping the boat's systems running smoothly—he's equally adept at designing a database, plumbing a boat toilet, and wiring an electrical panel. Fun Facts: he has to medicate to prevent seasickness, loves extreme weather, and stands out like a sore thumb in Central America.

**Tanya, First Mate and Ship's Cook.** The impulsive idea man, extroverted family ambassador, and neurotic control freak. She may be afraid of everything but doesn't let it stop her from living a full and exciting life. She loves planning trips, taking the night watch on passages, and is in charge of setting the anchor or picking up a mooring. She loves meeting new people, serves as French/Spanish interpreter when necessary, and knows how to find things in a new place. Fun Facts: she plays ukulele, reads voraciously, and likes to kayak.

**ELI, SECOND MATE.** The firstborn son, a frustrated perfectionist, a lover of the great outdoors, and a wordsmith lovingly known as Captain Vocabulary. He's in charge when Jay and Tanya are off the boat and helps stand watch at night on long passages. He's the one who goes up the mast when the need arises. Fun Facts: he loves to freedive, plays D&D, and is working on a private pilot's license.

**AARON, SECOND ENGINEER.** A Mr. Fixit, he loves tools, can talk to anyone with his charismatic personality, but can sometimes be a bit of a prima donna. He helps with boat projects like installing a water heater or changing the oil in the engines. His motion sickness limits his abilities on passages, but he's capable of piloting the boat in coastal waters. Fun Facts: he plays electric guitar, rebuilt his first carburetor at age seven, and knows almost everything about WWII tanks.

**SARAH, QUARTERMASTER.** A creative genius, who likes to draw and can play several musical instruments, and has a ready wit, though you might not know it because she's also a bit of a hermit. She helps with docking and anchoring, knows where to find anything on the boat, and enjoys sailing in small sailboats. Fun Facts: she's excellent at using just the right movie quotes to fit a conversation, is fluent in Spanish, and bakes the best cookies.

**SAM, ABLE SEAMAN.** He's got an indomitable spirit and the ability to charm animals and small children, yet somehow most often shows us his spastic clown persona. While sailing, he stands by to help wherever needed, and he likes to take morning watches on passage. He's the fisherman of the family. Fun Facts: he's a frustrated percussionist, a drummer without a proper drum set who taps on anything that resonates, and he can solve the Rubik's

Cube in thirty-two seconds, juggle, and touch his tongue to his nose (though not all at the same time).

**RACHEL, MIDSHIPMAN.** The youngest, born after we moved aboard *Take Two*, precocious and wise beyond her years, empathetic and imaginative, and possessing a flair for the dramatic that comes with the downside of a quick temper. She sleeps in a single bunk we built for her amidships and loves to sit in the captain's chair on passages. She is learning to pilot the dinghy and loves to help in the galley. Fun Facts: she adores animals, has a big singing voice for a small person, and can recycle anything for use as a toy.

*TAKE TWO*, **CUSTOM WOODEN SAILING CATAMARAN.** Our boat is more than just a vehicle that gets us from point A to point B; she is a part of our family. We love and care for her, and she, in turn, shelters and protects us. She was designed by Dirk Kremer and built at the Waarschip yard in Bouwjaar, Netherlands in 1991, the year Jay and I rode the school bus together in high school. She is forty-eight feet long, twenty-six feet wide, and has a draft of four feet. Her typical cruising speed is eight knots, but she's capable of double digits in a brisk wind. She was cold molded, made of cedar and layers of marine plywood and epoxy, with a fiberglass skin below the waterline. Her keels and cross beam are solid mahogany. We are her fourth owners, having purchased her for less than $200,000 in Fort Lauderdale in April of 2008. She has crossed an ocean, spent a few years as a charter boat in the Virgin Islands, survived several hurricanes and once even circumnavigated one. She's been our full-time home since August of 2009.

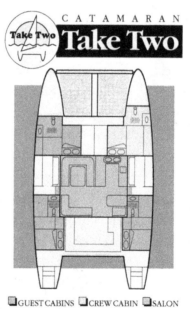

■GUEST CABINS  ■CREW CABIN  ■SALON

# PROLOGUE: STAYING AFLOAT
## THE WORST AND THE BEST

*May 2016. I am at the helm, the only crew still standing. The captain is wedged in a corner of the cockpit trying to nap. The others are lying prone, sleeping where they fell, some outside in the cockpit, others on the settees inside, and one, half naked, on the salon floor. If there were a soundtrack for this day, it would include crashing waves, wind whistling in the rigging, the drone of a diesel engine, crewmen moaning and groaning, and the sound of someone throwing up at the rail. The wind is wild, whipping my hair around and chapping my face. We are pounding into six-to-eight-foot seas, directly upwind, sails furled and both engines running. Occasionally, I get hit in the face with salt spray from the bows burying themselves in a big, green wave.*

*It is the kind of day people imagine when I tell them I live on a sailboat and they stare at me with an odd mixture of horror and admiration on their features. Perhaps they are thinking of the fisherman in his yellow rain slicker on the Gorton's fish sticks package. Well, sometimes it is like that, but only for a day or two out of the year. Sometimes, believe it or not, life at sea can even be boring. But usually, like this day, it is a combination of highs and lows, the highs often being better than you can imagine, and the lows, worse.*

*We are on a rhumb line between the east side of Puerto Rico and a small island in the Spanish Virgins, Vieques. The U.S. government once used Vieques for target practice, and despite its now being a vacation destination with beach resorts, there are still parts of the island that are off-limits due to unexploded ordnance. We are here in the middle of a churning sea because it was the best weather that we could see in the forecast for making our way east to the Virgin Islands. It is late in the season, May already, and we need to be in Grenada before hurricane season gets cranking. It's been a rough year for leaving, our intended departure date slipping from January to March because of Jay's work schedule and the numerous cold fronts and disagreeable conditions preventing our crossing of the Gulf Stream.*

*We passed up a month of cruising in the Out Islands of the Bahamas with good friends on* Ally Cat *in order to take advantage of a few days of calm weather to head east, and the last cold front of the season to push us south into the Caribbean. Though we're excited by what lies ahead, we are still feeling this disappointment. We had been trying to meet up with Kimberly and Michael and their daughter, Ally, for months, slowly heading south as they headed north, our paths crossing as they had twice before, in Washington D.C. and Fort Pierce, Florida. As it turned out, we had only three days together in George Town, Exuma in the Bahamas to catch up. We made the most of it, with a dinner together of freshly caught Mahi tacos, a provisioning day with two other boat moms, a cruiser's open-mic music night, and a beach bonfire. The last day, Kimberly bequeathed to me her notes from their year in the Caribbean, notes that I will cherish and use extensively.*

*One of her recommendations was Bio Bay, or Bahia Mosquito, in Vieques, a naturally occurring phenomenon where bioluminescent plankton exist in impressive concentrations in a closed bay and cause anything that passes through the water to glow and sparkle. I was enchanted by the idea of anchoring our boat at the entrance to the bay and taking our kayaks in on a dark night to give my kids a magical experience. I became obsessed with this idea—so driven, in fact, that when it was time to leave Puerto Rico, I insisted we make the stop in Vieques instead of going straight to St. Thomas, which might have provided a better wind angle for sailing. And now I am paying for it and exacting a price from my crew as well.*

*Guilty questions circle my head like seagulls after a potato chip. "Will this be one of those times when we all suffer for nothing? Like those other times when I have an idea and drag everyone along and it turns out to be a costly disappointment? Will we even be able to anchor at the mouth of the bay with the wind and waves from this direction?" I have six hours of bashing to think about this, while our little boat icon creeps across the screen of our chart plotter more slowly than I could ride a bike. I say a small, selfish prayer that it will all be worth it.*

*I have seen no other boats since we left this morning with our French counterpart—a boat called* Dingo D'Iles *(crazy for islands), another large catamaran with five kids aboard. They are long gone, heading to the British Virgin Islands. This is another disappointment, as we would like to have spent more time with them. We have never met another family with five kids aboard, and they had two teenagers too. We overlapped by only a few days at Palmas Del Mar, just long enough to hang out in the laundry room while catching up on the wash and to share Rachel's birthday with their three little girls. But they are on a schedule to get to Martinique by a certain date, and we are not. There is always the chance that we may run into them later.*

*Vieques grows incrementally larger on the horizon, as the mountains of Puerto Rico vanish behind us. The only redeeming qualities about this day are that it is not raining, and that we'll arrive before dark. I console myself, as I often do, by reminding myself* that it could always be worse. By midafternoon, we are running along the coast, looking for a place to anchor the boat. The captain looks dubious. The opening to Mosquito Bay looks too narrow and the bay itself too shallow for us to get inside, and the water is too rough to stay outside. I can hear him thinking about his bailout plan and calculating arrival time in St. Thomas. I cannot accept defeat so easily. Perhaps, I suggest, we could just do a drive-by and see whether it's doable.

So we creep in around a point, in whose lee lies a perfect little isolated palm-tree-studded beach, and inch toward the entrance to the bay. Suddenly, as if by magic, the wind and waves disappear and a mangrove-lined channel opens up just beyond a wide, shallow bight. We drop the anchor, fall back to see if we like the placement, and in classic Take Two style we reanchor. It's perfect. The captain agrees to give it a go, but we will only stay one night, so this is our only chance.

Everyone is moving again, like the waking dead, looking rumpled and groggy. "Where are we?" is the repeated question. And now that we are out of the wind, it's hot. And at the mouth of Mosquito Bay, Sarah points out, it might be a buggy night. But I remain optimistic. Yes, it might be hot and buggy, but we're in a safe place and, barring rain, we have a chance to go do something rare and interesting. Jay and I do an advance recon by dinghy to see how far we have to paddle, and what the bay looks like. We decide that I'll kayak with the big kids, and he will take the dinghy as a support vessel with our youngest crewmember, Rachel, who just turned five.

We make a quick dinner and drop the kayaks in the water. The sun sinks into the sea and stars begin to wink in the darkening sky. It is a moonless night, ideal for our purpose. We paddle down the long, serpentine entrance in the dark. There are a few sparkles in the water, but nothing we haven't seen before. A fish darts away from the bow of my kayak, and I see a streak of neon green. Then the creek widens into a bay, something we feel more than see. The farther in we get, the *brighter the swirls our paddles make in the water, until the water is unmistakably glowing. Fish dash in every direction leaving fiery trails like comets, the paddles come out dripping diamonds of light, and we leave radiant wakes behind us. The kids are all thinking the same thing, and finally someone says it aloud: "Can we jump in?" If it weren't so dark, Jay and I would exchange a parental glance. We had read that a girl was bitten by a shark in this bay a year ago, and we instinctively know that swimming in a warm, shallow bay at night is a bad idea. But we say yes, anyway. It's irresistible—a chance to swim in liquid light. Our friends on a boat called* Jalapeño *said it was not to be missed—they went so far as to dare our kids to swim here if they ever got the chance.*

*Our fearless firstborn jumps in first. His whole body is luminous. His hair is on fire with glints of green. One by one—even our timid five-year-old, who leaps in fully clothed—we all immerse ourselves in what looks like radioactive liquid. Our hands and arms come out of the water scintillating like we're wearing sequined gloves. The experience is thrilling, incomparable to anything we've seen or done. A kayak tour group emerges from a clump of mangroves and we have surely disturbed their quiet evening expedition with our riot of sound and light. We hop back in the kayaks after a while and play paddle tag, using the glistening trails to chase each other through the dark. This is what that awful day at sea was for; it has made all the discomfort worthwhile, and I am quite literally glowing with happiness. As we paddle out of the bay, the brightness fades, the streaks turning to mere sparkles again, and we head back for a freshwater rinse and bed.*

*Tomorrow, we'll weigh anchor and head back out to sea. The waves will still be there, but hopefully we'll have a better wind angle for sailing to St. Thomas. We'll be sailing past Culebrita, with its famous jacuzzis, a series of natural rocky pools on an island wildlife refuge. Our good friends on* Abby Singer *are anchored there, but time and weather do not allow for another stop, so we'll have to catch up with them farther down island. So goes the life afloat.*

---

Sometimes we measure success on the boat by the absence of failure—nothing broke! Nothing leaked! No one got seasick today! Sometimes sailing looks merely like not sinking. There are glorious, wonderful, sparkling days, but they stand out in memory like an oasis in a desert of rough passages. Staying afloat acknowledges the hope-amidst-hardship of the sailing life. If it's so hard, one might ask, why do we do it? Because despite the unpredictable and sometimes unpleasant nature of boating, the beauty, joy, and freedom we experience in nature, the sense of accomplishment we feel when we overcome a challenge, and the memories we make as a family while traveling make it all worth it.

Disappointment is a normal part of life on Planet Ocean. Our life and path are often dictated by things outside our control, like the weather, Jay's work, or things that break unexpectedly. We may yearn to go somewhere but be unable to get there because it's the wrong time of year, or the wind is blowing the wrong direction or speed. While we love to go off the beaten path, we can't stay very long and keep the paychecks coming. This is partly why we have not crossed an ocean yet, and why we waited so long to make the jump to the Caribbean. We were waiting for the technology to catch up with our dream so that Jay could work from the boat wherever it was anchored. The tradeoff is that we get to live this way, instead of saving up for ten years so we can take a trip.

Then there are the things we can control. Every time we say "yes" to one thing, we have to say "no" to a thousand others, some of which may have been better than the one we chose. Often, we pray through a decision and choose a counterintuitive path whose purpose is only revealed later. But there is no loss without some gain, and when we miss a time with old friends, for example, we have an opportunity to make new ones.

Our lost month in the Bahamas with *Ally Cat* was later spent in the Virgin Islands cruising with *Abby Singer*. The weeks we might have spent with them in Culebra were used to earn income and tour Puerto Rico by car. A rough day at sea yielded a memorable night in a phosphorescent bay. Choosing to continue feeling disappointment about lost joys keeps us from experiencing new ones. We just need to stay afloat during the hard times so that we are ready when good times come again. This is one of the chief lessons we have learned from life on a boat, though not the first.

# 1

## ROCKING THE BOAT
### BIG DREAMS

*February 1994. I am sitting cross-legged on the cold floor in the basement of the computer lab, an ivy-covered stone building in the middle of my college campus, a telephone cupped to my ear. The voice on the other end sounds deceptively close.*

"*Where should we meet tonight?*"
"*How about Greece?*"
"*On a boat?*"
"*Of course...sailing naked in Greece. See you there?*"
"*Goodnight...I love you.*"
"*I love you too.*"
"*See you in Greece.*"

*I hang up and reluctantly return to my English paper. It's two in the morning and I'll be headed back to my dorm in a few minutes to get some sleep before class.*

*More than a thousand miles separates Jay and me, and the bond we first forged as high-school sweethearts is held together by a coil of thin wire stretched between a payphone in the computer lab of Middlebury College in Vermont and a dorm room at Georgia Tech in Atlanta. While I'm working on papers about Chaucer, Milton, and*

*Shakespeare, Jay is studying for calculus tests and learning about fractals. When the loneliness gets to be too much, I slink to the basement in the wee hours and call him. We are lovesick and sleep deprived, and the conversations follow a predictable pattern. We are crazy enough to think that concentrating all our energy on the same dream as we fall asleep will somehow result in our meeting there, in Greece or some other exotic location, sailing. Sharing this dream keeps us together, though sometimes it feels like the dream is all we have in common.*

⌇

We were the poster couple for opposites attract. Our high school psychology class teacher, Mr. Stump, evaluated our personality tests, looked at us, and said, "It'll never work." Perhaps we took this as a challenge. I was a practicing Christian; Jay was a staunch atheist. I have an outgoing, talkative, and friendly personality; Jay is quiet, solitary, and introspective. I tend to be goal driven; Jay is happy to go with the flow. I am a writer; he is a computer geek.

Jay was steadfast. He fell in love with one girl. He wanted nothing else than to marry her and make a life together, someday. I was easily distractible. On the one hand, Jay was my first and most intense love experience. On the other, we were so different —on opposite ends of the personality scale, you might say—that I worried we weren't compatible long term. I wasn't sure what I wanted yet. And he could outwait anyone, while I grew lonely and impatient. I wondered if anyone else was like him. I thought life was like an ice cream shop, so I tried other flavors, only to find the others' kisses were lacking, and instead of forgetting the favorite, I simply longed for it more than ever.

High school is a time in life where all dreams are still possible.

We had this crazy idea: that we would buy a small boat and sail away. So why did it take us more than a decade to make it happen? We had been raised inside a system in which one's life is booked like a passage on a big ship: there are rules, procedures, and only one accepted way to get from the port of departure to the destination. I don't remember boarding, but I was certainly warned not to jump off! So Jay and I steamed as fast as we could in the wrong direction, for a little while at least. And, in a way, that experience made the adventure that came later possible: we earned the money we needed to escape the system, and we learned what we did not want, a valuable part of finding one's path in life. But nothing can rock a boat like a dream that won't go away.

Anyone who has stepped into a dinghy or a canoe—hopefully with care—knows what it feels like to experience Newton's Laws of Motion firsthand. A misstep, or too much weight placed on either side of the center, and one can be thrown off balance. Overcorrection sometimes results in an accidental swim. On a bigger boat, the rocking motion can be disorienting—the information coming to your brain from the eyes, the inner ear, and the muscles of your feet and legs may be sending conflicting messages. For some people, this causes sleepiness, headaches, and nausea, and in some cases, incapacitates them completely. However small or large the boat, the rocking is uncomfortable.

Perhaps that's why we are told our whole lives, "Don't rock the boat." At best, it means keep the peace, but at worst, it means maintain the status quo. It sends the message, "Don't do anything unpredictable, uncertain, unsafe—don't take any unnecessary risks." But a life without risk is a life without adventure. Stepping into a boat is synonymous with launching into the unknown: exploring, dreaming, and discovering. It comes with the inherent danger of being injured, drowning, getting lost, infighting, and coming into contact with creatures from the deep, storms, and pirates. But it also holds the rewards of beauty, freedom, joy, community, and a sense of accomplishment.

The realization of a dream takes three things: a clear destination, a thousand small steps, and an endless supply of sheer stubborn determination. The destination, or big idea, should be something you can picture clearly in your mind in its completed form. It's something you can describe, in gory detail, and talk about like it's a foregone conclusion. My friend Todd (who traveled on a boat called *Paisley* with his family) says that to make something happen, you have to make a verbal commitment. The more you talk about it, the more likely you will be to make it happen. (He also says that on a scale of "bored to dead" it's better to stay closer to dead, but I digress.) Of course, talk is only the beginning—to take it to the next level, actions must follow words.

The steps are taken one day at a time, every small decision steering you toward the idea. Steps that lead away from the end goal, no matter how appealing or reasonable, often have to be abandoned. This is difficult and requires courage—you may have to distance yourself from people who discourage you and even the best plans might be discarded if they don't serve to advance the goal. (Logically, it follows that walking in the wrong direction means backtracking later, and a longer route to the destination.)

And then there are the obstacles that are continually thrown in the path, difficult or seemingly impossible circumstances that prevent you from moving forward. Some people hit these inevitable obstacles and assume it is a sign that they're on the wrong track. Some people are discouraged and give up on their dream. Others, like soldiers in boot camp, hit the obstacles and climb over them, refusing to take no for an answer. These are the people who make an idea into reality. They are stubborn, determined, and sometimes ruthless, the salmon swimming upstream, like me, or like Jay, the boulders in the middle of the river, steady, strong, and immovable types. In Randy Pausch's *The Last Lecture*, he says that the "brick walls are there for a reason. The brick walls aren't there to keep us out. The brick walls are there to give us a chance to show how badly we want something."

Few parents tell their kids that their dreams are stupid or crazy: kids are expected to dream big. Adulthood is when you become responsible, hang your unrealistic dreams in the back of the closet, and settle into *real* life. Maybe retirees are allowed to rekindle the dreams of youth and pursue them, but only after they have had a productive career and played their part as a cog in the machine. But we were just seventeen—two kids, walking on the city docks in Naples, Florida, holding hands and talking about what it would be like to ditch it all and sail around the world. This was Jay's idea—he was the sailor—and it seemed romantic, if a little unrealistic.

My ideas were more practical, but also contained travel. I would go to college, study French, spend a semester in Paris, get my degree, and pursue a teaching career. I grew up enjoying summer road trips with my family (as in "take five people who can't get along in a house and jam them in a car for a few weeks and drive across the country.") Teaching would enable me to keep summers free to travel.

So, not knowing what else to do, we went to college, which, in our case, meant dating long-distance.

Both of us were independently seeking answers to the hard questions—the meaning of "life, the universe, and everything," as Douglas Adams puts it. That's what you do when you're growing up. Growing apart would have been easy as we were going through this process separately. At this stage in life, four years seems like an eternity. I am still amazed that we were not torn to shreds in the vortex of early adulthood.

We survived holiday to holiday. We couldn't wait to be together, but we were having new life experiences and meeting new people in two very different spheres. Whenever we saw each other again, once the initial euphoria wore off, it took a day or two to get reacquainted. After the awkward beginning, we would be comfortable like old shoes again. At the end of the holiday, I would go back to my 200-year-old dormitory in the country, my writing, and my friends, and he to his apartment in the city, his cat, and his computer. It was a fragile existence.

By our junior year, we were used to the idea of dating long distance. We didn't like it, but we had made it this far and we figured if we could make it until the end of college, we would be able to handle anything married life could throw at us, which, as foolish as it sounds, has held true. That was the year I went to Paris to study abroad for the fall semester. It was a test of perseverance and faith in each other, one which we passed by the skin of our teeth. I was an ocean away. The French Poste went on strike; then the Paris Metro went on strike. Once, three weeks went by with nothing in the mail. The difference in time zones and dollar-a-minute phone rates made talking almost impossible. The only way to communicate was via the newfangled electronic mail, but to get to a place where I could use it, I had to walk thirty minutes in wintry weather to my campus, where I had a limited amount of time to download any incoming mail and upload the email I had written. It wasn't great, but it was better than nothing.

Jay, meanwhile, had dropped out of Georgia Tech, and had rented my empty bedroom in my parents' house in Florida, trying to make it on his own. He was a young entrepreneur computer genius, working eighteen-hour days on a project (fueled by caffeine and adrenaline) that he hoped would make him rich and let him retire by age thirty. And he was having an existential crisis in his spare time that ultimately led him to a budding faith in a Creator. Though he was rarely home, my parents had grown to like him, and we could see the proverbial light at the end of the tunnel. After nearly six months apart, the hardest wait was the ten-hour flight from Paris to Miami. We had not only survived, but grown even closer. A few months later, we were engaged.

In the whole pie of life, these four years are but a small slice. From these glib little paragraphs, it might be hard to sense our long-suffering and heartache. The more fun we had when we were together, the more miserable we were at parting. There was always uncertainty because youth is a precarious time, and things can change quickly, destroying the best laid plans. We developed coping mechanisms. We talked on the phone. We laughed. We cried. And we wrote letters, hundreds of letters: letters scented with perfume to remember special occasions, booby-trapped letters which dropped glitter on one's lap during class, sappy, sad, poetic, and sometimes frustrated letters. We illustrated them, we cried on them, we spilled things on them. They are now treasured keepsakes, tied in bundles in a box in an air-conditioned storage unit along with wedding pictures and baby books. We open the box and peek at them occasionally, pull one at random out of the stack: they drip with angst and longing, hope and doubt; they burst with dreams of Someday. Those letters kept us afloat.

We developed a special knack back then that has served us well throughout our life together. We were able to visualize that elusive Someday, and then work toward it as if it actually existed. It is not much different than trying to meet in Greece on a sailboat while sleeping. We thought the same thoughts day after day. And though we had absolutely no idea how to achieve it, that's when we first decided to sail away.

As teenagers, holding hands and walking the docks, we would look at boats and talk about places we wanted to see and things we wanted to do during our limited time on Earth. But we had no idea how to turn those dreams into reality. After I graduated from college in 1997, we got married and moved to suburban Atlanta. I taught elementary school while Jay pursued a career in computers. We had an apartment, two cars, two cats, and fifty thousand dollars of debt. After a couple of years, we bought a house with a white picket fence and some furniture. Around our fourth wedding anniversary, we had a baby boy whom we named Eli. A year later, we had a second son, Aaron. When we got pregnant with a third child, a girl we would name Sarah, we bought a minivan with room for three car seats. Jay made enough money that I could stay home with the kids, and he drove a sports car on the weekends. We were living the American Dream.

But deep down, we felt unsettled. In the house with the white picket fence, we still talked about our idea, but it seemed like a mirage—something that shimmered on the horizon but vanished if we got too close. Instead, Jay got his sailing fix by racing sailboats on Lake Lanier on Wednesday nights. We would visit family in Florida and wax nostalgic when we saw boats or stood on a beach and looked at the endless ocean. We subscribed to no fewer than five sailing magazines, our favorite being *Cruising World*. We read anything we could get our hands on about living and traveling on a sailboat, books with titles like *Survive the Savage Seas, How to Sail Around the World,* and *All In the Same Boat*. We kept the flame of that small idea from going out, and it kept us warm for more than a decade. When the opportunity arose to leave our predictable life, we would be ready to step into the rocking boat.

We had gotten a glimpse of what it could be like on our first sailing trip together as a married couple. Jay's dad and stepmom, Al and Mary, had invited us to join them on a weeklong cruise to the Florida Keys on their Prout Quest 31, *Double Entry*. This is the boat on which Jay had sailed as a child, the boat on which he was working the summer we fell in love, when he sanded and painted the bottom in a boatyard and I brought lunch, which we shared under the shade of the bridge deck. That trip had offered me the chance to become acquainted with the seafaring lifestyle of Jay's childhood, and to get a taste of it for myself.

I had started the trip with trepidation; we spent the first night out of Naples pitching and rocking through the dark and the rain toward the Dry Tortugas. Jay got seasick and threw up, then I got queasy and joined him at the rails. I called it a bad case of sympathy seasickness—the only other time in my life that I had been seasick was on a dive boat in the Florida Keys when I was a teenager—with Jay, when we threw up overboard side by side (how romantic). I would think that his seasickness would be the cure for his sea fever, but evidently not.

But once we had reached the Dry Tortugas, a remote National Park with crystalline waters and pristine coral reefs, I began to see the charm. Fort Jefferson, a hexagonal, Civil War era brick fortification built on Garden Key, provided a chance to go ashore, walk around, and learn some obscure history. We sat on deck talking into the night, enjoying the quiet sounds of water lapping at the hulls. During the day, we swam off the boat, relaxed in the cockpit, and read books. After a couple of days, we sailed to the Marquesas and snorkeled. When we pulled up the anchor, we found a small octopus holding on to it for dear life. Our days in Key West were filled with good meals and old stories.

The trouble started as we were leaving Key West, when I bumped into a fragile fitting and knocked the fuel line off the tank, starving the outboard motor until it sputtered to a stop. Though Jay and his dad jury-rigged the fitting and restarted the motor, they couldn't get it into gear. At least that part was not my fault. The result was a forced all-night sail in a light breeze from Key West to Naples Bay, where we got a tow back to the dock. But it was so much more than that.

As we were sailing back to Naples, the sun set and the stars blinked on one by one, like little night lights. Jay's dad was on watch. Mary had gone to bed already, and Jay asked if I was coming down below with him. I felt so awake and alert, electrified by the beautiful evening, that I decided to sit up a while with Al and enjoy the night sailing. In the water, tiny phosphorescent lights mirrored those in the sky. I sat up all night, enthralled, not wanting to miss a moment, wishing we were sailing away from civilization and not back toward it, toward our life in Atlanta—our apartment, our jobs, our bills. I found myself wondering, *Why do we have to go back to that kind of life? Why can't we just go sailing instead?*

Jay and I talked it all over during the long drive back to Atlanta, the stories from his childhood, the new things I had discovered about a person I thought I knew well, about how fun it was. I told him about my magical night at sea and asked him that boat-rocking question: "Why can't we keep doing this forever?"

## 2

## UNCHARTED WATERS
### A LEAP OF FAITH

*October 2003. We sit down at the dining room table with a contract between us. The boys are napping on this quiet Sunday afternoon and we have a decision to make and papers to sign. Friday, we met a real estate agent in the Ashford Dunwoody neighborhood and he showed us a suburban split-level house with a nice backyard, good schools nearby, and convenient shopping just around the corner. Though the starter home we live in still has plenty of room for our growing family and we adore it as one does a first love, the neighborhood seems to be taking a turn for the worse and we worry about the schools not meeting our standards and property values plummeting. Jay is spending as many as three hours each day sitting in traffic on Interstate 285, and this move would take us closer to where he works.*

*All our friends, also young professionals, are moving out of their starter homes and into bigger and better houses outside the perimeter and buying luxury vehicles. Neighborhoods of McMansions are sprouting up like an invasive species all around Atlanta, forests of live oak trees replaced by foundations and wood frames, squares of sod, and saplings tied to stakes to keep them upright. Like having babies and buying a minivan, it seems like the logical next step.*

*But something is amiss. Call it a gut feeling if you will, but both of us are reluctant to sign the paperwork. Neither one of us wants to rain on the parade, but we quickly discover that we have independently come to the same conclusion: we should not buy this house. We are praying people, but not mystics—we didn't get a sign from God or hear an inner voice—we just have unanswerable questions and a feeling of unease. If this is logically the right decision, why do we feel so uncertain? Is this an important fork in the road—will buying this house entrench us in suburban Atlanta and take us further from a dream of sailing away? One of us irrationally asks why we need to buy a house at all. Couldn't we just move onto a boat instead? (Never mind that we are about to have three children under the age of three.)*

*We laugh at this notion, but all the same, we call the real estate agent and tell him we've changed our minds.*

"The house we didn't buy" became the turning point, the moment when we departed from the charted course and began to navigate by a different set of expectations. It was just as exhilarating and nerve-wracking as sailing through unknown waters. Looking back, it seems miraculous that we escaped suburbia at all. So many circumstances coincided to make our path clear that we have come to believe that we must have had Divine Help. Without faith in unseen possibilities, we would probably still be living safely in a suburban split-level in Atlanta around the corner from a Starbucks. But faith in a future dream and hope that God will guide us are only navigational tools, without practical skills and momentum, we would never have headed in the right direction.

We often perceive dreamers as people who lack practicality,

but those who achieve their dreams must think of the minutiae. Without money, for example, how could one ever afford to sail off into the sunset? When we were first married, someone gave us a copy of Larry Burkett's *Complete Financial Guide for Young Couples*, a book that helped us set our priorities straight from the beginning. That, paired with Richard Swenson's *Margin: Restoring Emotional, Physical, Financial, and Time Reserves to Overloaded Lives*, gave us the practical underpinnings that would make buying a boat and moving aboard with our family possible.

The first thing we decided was that if we had children, I would stay home with them, and if we wanted to go sailing instead of living in a suburban neighborhood, I would have to think about how to educate them on a boat. So, while we couldn't depend on my salary for our living long-term, I would need to use my teaching job to hone my communication and organizational skills, familiarizing myself with curriculum and thinking about how I would approach homeschooling. And Jay would have to find a job that could either be done remotely or create residual income. For the first few years of our marriage, he had a brick-and-mortar job at a consulting firm, often traveling to see clients in person, but he was building the skills and reputation upon which an independent practice and remote office would depend.

The second principle we put into place was digging ourselves out from under our debt so that we could live within our means. If we weren't free financially, then we would never be able to set off on an adventure. Debt keeps people enslaved so that they can't pursue a calling other than working day in day out to pay the interest on loans for things that, for the most part, do not bring lasting joy. So we spent the first two years of our married life in a small, dingy apartment, with nothing more than Jay's black leather bachelor sofa, a kitchen table and four chairs given to us by his mom, a mattress and box springs which were a wedding gift from my parents, some plastic storage bins, and two black and white kittens, Sugar and Spice. And we were happy. We used money that we might have spent on furniture or vacations to pay down high interest credit card debt. Then, as each debt was cancelled, we tackled the next one: car loans, school loans, money borrowed from family. Between Jay's and my salary, our frugal living, and the forgiveness offered by family members and Uncle Sam (who cancelled some of my college debt because I taught school in an impoverished area), we were able to bail ourselves out in a little over two years.

The lower interest school loans took a little longer, and we took on a mortgage when we bought our first house, which we viewed as an investment, but besides those payments, we were free. When I quit teaching to stay at home, we lost the gravy of my salary, but not the meat and potatoes we needed to live on. Better yet, we had put a third principle to work: we learned to partition our income by percentages; giving, saving, investing, housing, and spending each had a slice of the pie, and we learned to live on less than what we made. Figuring out the finances is one example of a practical step we took to make our dream possible, though not the only one.

Sometimes deciding what you *do not want* is just as important as deciding what you *do want*. We were not unhappy in suburban Atlanta—after five years of marriage, we had moved up in the world, having relocated from our dumpy apartment to a ranch style, three-bedroom, two-bath home on a half acre of land. We planted a rose garden in the front yard, hung ferns on the porch, made the back deck and yard a comfortable place to hang out on weekends. We painted and wallpapered, bought furniture, hung art, and beautified our home—all the things young couples do. We drove nice cars. Jay had even acquired a Porsche Boxster; after the kids came along, it was our date car. We had a walk-in closet and dressed well. We went to fancy restaurants, concerts, and company parties. We took vacations. We joined a church full of people that looked just like us. Jay crewed for sailboat races on Lake Lanier and went to the driving range to practice his golf swing. I went to Middlebury alumni functions and volunteered for charities. By the time Eli and Aaron came along, Jay was well established as a computer consultant and made a good living. I was able to quit my teaching job, which I loved, to stay at home with our boys, a job I loved even more. I joined play groups with other stay-at-home moms, so we could chat while the toddlers played. Outwardly, at least, we were the typical yuppie couple.

But we weren't really satisfied. I should have been content—I had everything I had thought I wanted. I had spent my childhood moving from house to house every few years whenever my dad's job would change, so the suburban lifestyle provided stability and comfort. But the goal of keeping up with the Joneses wasn't enough for us. Our neighbors drove nicer cars, took fancier vacations, and planted prettier gardens. We weren't willing to go into debt to compete, and the act of acquiring more and larger stuff did not appeal to us. I found myself thinking, *So, this is the American Dream. What's all the hype about?* Life in suburbia lacked a sense of adventure.

Maybe *that* is why we did not buy the house in Ashford Dunwoody. In May of 2004, we brought our third child, Sarah, home from the hospital. Jay went away a few days later to interview for a job in Tampa, Florida. We had decided that the first step to pursuing a new life was to go back to Florida, the place where it all started, to be near the coast and our families. We had thought and prayed about the decision, and we felt like the circumstances confirmed that we were on the right track. We put our house up for sale and quickly had a contract and a closing date. Jay's mom and stepdad owned a condo on the beach near Clearwater and they decided to sell their Atlanta house and move to Florida permanently. We were even making plans to share a moving van.

But then, instead of falling into place, it all fell apart. The company that interviewed Jay said he was overqualified for the job. The buyer for the house failed to fulfill the contract and disappeared; the closing date came and went. Jay's parents packed up and moved into the condo on the beach. We were in a quandary. This was not the first, nor the last, time our faith was tested. At the heart of our belief in a caring God is this question: *Do we trust God to give us the desires of our hearts?* This is not some Cosmic Santa, but someone who knows what's best; who gives us what we desire *if* that's what's good for us, and, if not, replaces our desires with better ones. Since we can't see the future, we have to be patient and wait for God's plans and timing to reveal themselves. So, we learned how to wait.

We waited like sailors in uncharted waters, watching for a sign above or below the surface to tell us which direction to turn. Something was about to happen, but we weren't sure what. Would we run aground on our expectations or find a clear passage through all the options? We continued to pare down our belongings as if we were moving soon, though we didn't know where. We began to clean out the closets and basement, to donate or sell things to be ready in case the house sold, to simplify our lives before the need arose. This habit of believing something when the outward signs are not present would help us every step of the way. It is the essence of our faith—trusting in Someone who can see the next steps to help us move forward when we can't see. Hope in this case is not merely wishful thinking, but confidence that God is working out a Good Plan and that He'll let us know when the time is right. It is the spiritual equivalent of navigating in unknown waters without the aid of navigation equipment.

On a modern cruising boat, one navigates using an electronic chart plotter, a computer with digital charts that shows speed, course, GPS waypoints, and obstacles on the rhumb line—the straight path between point A and point B. A depth sounder uses sonar to show how deep or shallow the water is and may even have a digital display that paints a clear picture of bottom features. A wind indicator shows direction and acceleration of air currents and an autopilot can even be programmed to adjust course based on wind shift, underwater current, or compass heading to keep the boat going toward the destination. The use of the radar and AIS (Automatic Identification System) can offer a clear picture of what lies ahead—approaching thunderstorms or other ships in the area. Unlike the seafarers of old, sailors in recent times rely more heavily on electronics than on an old-fashioned paper chart, compass, barometer, log, plumb line, spyglass, and sextant—and the navigational skills that went with them.

But sometimes electronics fail, and sometimes a sailor finds himself in uncharted waters, areas where little or no information about depth or topography is available. In frequently traveled areas and near shipping lanes, hydrographers have created detailed, color coded, nautical charts—maps which show bottom depth and type, shoreline, inlets, islands, reefs, shoals, and navigation aids like lighthouses and channel markers. They may even show the best path between two points, good anchorages, and details like shipwrecks, rocks, and structures on shore. But sometimes accurate charts do not exist for remote areas where mapping the seabed is neither necessary nor financially feasible. Other times, there are shifting sand bars or areas of scattered coral heads that are not only uncharted, but *unchartable*. Then there are the cases where charts for an area prove to be outdated or completely inaccurate—without local knowledge or visual navigation skills, it would be easy to run aground.

When we left our house in suburban Atlanta, we also left the well-travelled shipping lanes of life, and by choosing an alternative lifestyle, we were venturing into uncharted waters. We would need intuition like that of the ancient mariners to find our way forward. On a boat, small errors of navigation can mean the difference between floating and sinking, so diligence and caution are required to avoid disaster. In such situations, the captain must slow down, take in and interpret clues from the environment and the tools at his disposal, and post a lookout on the bow to communicate when there are obstacles so the course can be altered quickly. We didn't know it at the time, but Jay and I were developing this kind of partnership and communication, trust and intuition, that would prepare us for life on the boat as much as it helped us leave the city.

The first week of August, around the time of our seventh wedding anniversary, we moved to a house in Clearwater, Florida. It was situated in a nice little neighborhood full of sherbet-colored homes with tiled roofs and pretty gardens. It had four bedrooms, a large, open living space with sliding glass doors that looked out onto the patio and swimming pool. It was perfect for a growing family, and got us one step closer to the water, to our extended family support system, and to our dream of sailing away. We had made a five-year plan—to downsize, to move to Florida, to buy a small boat for learning the ropes, to sell everything, to buy a large boat and move aboard, and to head toward the islands. The Clearwater house was just temporary—a port-of-call on a longer journey. And how we found it was Providential. I usually just skim over this whole part of the story—when someone asks me how we did it, I say, "we moved back to Florida, bought a boat, sold everything, and sailed away." But if they want details, this is what I tell them.

We thought we had made a good exit plan to extract ourselves from Atlanta, but when it fell through, a better plan revealed itself. Jay's boss found his profile on a head-hunting website and asked him about it. Jay was honest—that we were looking for a way to move back to Florida. Instead of being offended, he offered Jay an opportunity to open a Tampa branch of the consulting firm. The company would pay for our move and whatever was needed to set up a home office. Around the same time, a home inspector knocked on our door one afternoon while the children were napping. He said he was finalizing documents so we could close on the house the next week. I was baffled. I invited the man to have a seat on our front porch while I called Jay to see if I should let him in.

"Hi, honey. There's a guy at the door who says he's a home inspector for our buyer. Do you know anything about this?"

"What? No, our contract fell through…what buyer?"

"I don't know. He says we're closing next week. Is he at the wrong address?"

"Can you hold on a minute? I'll call our agent and see what she says…" After a long pause, he responded, "Well, she says it's a long shot, but she thinks it's the same buyer. Let him in, just in case. If it makes you uncomfortable, you can keep me on the phone while he's there."

So I let him in. He inspected the house without waking the children, handed me a business card, and left.

I had planned a Fourth of July trip with the kids to visit grandparents the following week, so we signed power of attorney papers just in case Jay had to deal with the sale of our house. After I left for Florida, the mystery buyers showed up, bought our house, and gave us a month to vacate. I was in Clearwater visiting

Jay's mom and stepdad on the beach. When I got the call that we had closed on the house, I suddenly had a weekend to find us a place to live.

Jay's mom recommended a local real estate agent who could show me around. She watched the boys, then aged two and one, while I carried a nursing baby all around town looking at houses. We walked through more than ten houses in two days, and still I hadn't found one that fit our needs. They were either too expensive or needed too much work. Once the agent got a feel for what we wanted, he said he knew the perfect house for us, but he would have to make a call before we could get in to see it, as it was not on the market yet. His sister had just gotten married, and the couple was combining households. His new brother-in-law's house would soon go up for sale, so he called to see if we could take a peek.

The minute I walked in the door, I knew that this was the right place for us. It was the least expensive and most renovated house I had seen, and in a neighborhood we had been eyeing. We made an offer, the owner accepted, and the house was sold without ever having had so much as a sign in the yard. Jay didn't lay eyes on the house until he flew down for the inspection two weeks later. At the end of the month, Jay's parents were coincidentally going on vacation the week we were relocating, so we camped out in their condo while the movers packed up our Atlanta house and drove the truck to Florida. With three kids under three, things went much more smoothly than I could have planned.

Five years later, we called that same real estate agent, and he sold the same house in a matter of weeks, during the housing crash of 2009, when it seemed like there were for sale signs on every other lawn. By then, we had followed the rest of the steps in our five-year plan and had just moved aboard our boat, *Take Two*, so the house was furnished but uninhabited, perfectly poised for a quick sale. The only reminders of that house are the port-of-call decal on the back of our catamaran, and the good friends we made while living there, with whom we still keep in touch.

With a little faith, the seemingly impossible becomes easy. This adage works for both navigating on the water and in life. We have learned that when we set off into unknown territory and things seem uncertain, sometimes we just need to back off and wait for clarity. If we are entering an area of unknown depth or unmarked shoals, we move forward slowly, ready to stop or reverse if necessary. When making big life decisions, we must also be willing to change direction, rethink plans, and find a different approach, trusting that the best way will make itself clear. Navigating uncharted waters, whether literal or figurative, requires a combination of faith and action, intuition and skill, and, of course, trial and error.

# 3

## SINK OR SWIM
### SURVIVAL SKILLS

*February 2005. We wake to a typical Florida winter morning: refreshingly cool, breezy, and sunny. Tampa Bay sparkles like a sequined dress under a disco ball. I have been begging Jay to take me sailing for weeks, ever since he bought* Blue Bear, *a Ranger 22, a small, fast monohull which we plan to day sail. In his husbandly wisdom, he's been waiting for a perfect day, afraid that if I have a bad experience our whole plan could be sunk before it's left port. I'm standing in the cockpit, ready to go, the kids are at Jay's mom's for the afternoon, and Jay is untying the lines.*

*We putter out of the marina and past the Coast Guard station. Out on the bay, we head upwind to raise the sails. Jay has had a chance to work out some of the kinks before taking me along. We head upwind first, tacking our way out into the bay, zigzagging toward the Sunshine Skyway bridge. Jay literally shows me the ropes as I'm starting basically from scratch. "Pull that thing…and that other thing!" is about the limit of my sailing vocabulary, which I learned from watching* The Princess Bride. *He explains the difference between the jib and the main, a tack and a jibe, how to trim or ease the sheets, how to duck and move to the other side when the boom swings as we come about. The boat heels and we sit on the high side, bracing our feet against the opposite cockpit settee.*

*After an hour or so, we turn the boat around and head downwind, back toward the marina. Jay hands me the tiller, but I have no idea what I'm doing, like a kid thrown into the deep end of a pool. I feel the force of the water against the rudder, the wind pressing against the sail. It's like a living thing, this little boat. I understand for the first time why they give boats names and call them "she" instead of "it." With a few instructions, I manage to get the boat headed in the right direction. I am enchanted: it's even better than I imagined. How could it not be? No kids, sunshine, salty breeze, the boyish sailor at my side…it feels like we're teenagers again looking at an endless horizon. Anything is possible and it all starts today.*

⚓

All of our children learned to swim early. The house in Clearwater had a pool, and the boat we live on now has an ocean in the backyard, so it behooved us to give our kids the survival skills they would need on the water. We are not cruel—we never literally threw them in for a sink-or-swim lesson. But we did reach a point in their swimming practice after they showed enough proficiency, when we tossed them fully clothed into cold water so they would know what it feels like to swim for survival. We stood by, cheering them on while they swam for the ladder, a look of surprise on their faces. Lifesaving skills may be learned little by little in a safe environment, but the day they are tested may still come as a shock.

One might describe us as cautious adventurers—while we don't mind rocking the boat, we don't want anyone to drown. Our getaway plan involved a series of baby steps. (Remember Bill Murray's character overcoming his fears in the movie *What About*

*Bob?*) The first step was *move back to Florida*. The next step was *buy a small boat and learn to sail as a family*. The third was *charter a boat for a few weeks to try it out*, though we soon discovered that this kind of vacation can cost over ten thousand dollars and might not be fun with three toddlers in tow. Eventually, we would *buy a medium-sized boat for coastal cruises*. And, someday, when we were finally ready, we would *buy a boat and move aboard*.

A few months after we moved to Florida, we sold Jay's Porsche and bought *Blue Bear*, which we sailed on weekends in Tampa Bay. It was often a family affair. We had a bucket of LEGO bricks to keep the boys happy, though they turned the small cabin into a minefield. One-year-old Sarah often sat on Jay's lap wearing her colorful life jacket, the wind blowing her wisps of blond hair. The three kids remember napping in the V-berth, rolling over whenever we tacked. I fondly remember the sparkling afternoons on the bay, listening to music and seeing my husband happier than he had ever been. It was like watching a bird in flight, a creature doing what it was meant to do. He spent every free minute on the boat, fixing it or sailing it. I was supposed to be learning to sail in my spare time, but soon I was pregnant with a fourth child. I spent as much time as I could on the boat, though I soon became large and ungainly. Sam was born February of 2007 and we have a picture of him on *Blue Bear* at only a few weeks old.

In the marina where we kept *Blue Bear*, there was a Tayana 55 called *Katie Rose*, a huge fiberglass sailing monohull. It had a for sale sign on it, and we called the broker out of curiosity. When we climbed aboard for the first time, an overwhelming smell of old wood and diesel fuel met us in the companionway. All we could think was, *this would be a lot of work*. But over these doubts, I heard something else loud and clear, "This is your future." I actually heard a *voice*—not in my ears, but in my mind.

This thought was at once exciting and repugnant. It would be a huge departure from our step-by-step plan and put us on a liveaboard boat much sooner. But the boat had been sitting for five years, its owner having made it halfway across the Pacific before falling ill and returning to Florida. Health problems prevented him from sailing or maintaining the boat, and eventually he had to move to Montana to be near family. His son was selling the boat.

We could gather all our resources and pay cash for her, but we didn't know if we could afford to fix her, the old adage being, "A boat is a hole in the water where you throw money." Also, she had only three cabins with sleeping spaces for five, but we were now a family of six. We would need to do some major work to make it livable. It felt like biting off more than we could chew, so we waited and prayed about it. The asking price went down. We looked at the boat a second time and prayed about it again. The price went down again. Still we hesitated. Here, on the cusp of realizing our dream a lot earlier than we expected, we found ourselves reluctant and fearful.

One weekend, a retired couple came to see the boat. The next Saturday, while Jay was down at the marina working on a project on our boat, *Katie Rose* pulled away from the dock. He called me.

"You're never going to believe this. Remember when I told you about that couple that was looking at *Katie Rose*?"

"Yeah?"

"Well, they are heading out of the marina, on their way to the Chesapeake for a refit. I just talked to the broker on the dock."

"What???" I replied incredulously. "How is that even possible? They just saw the boat last week! What about all the repairs it needs? How are they going to make it to the Chesapeake?"

"I have no idea. He doesn't even think they're going to make it to Miami."

"But that is *our* boat! We know everything about that boat...I really thought we were going to buy her."

"Yeah. I'm pretty surprised. But maybe we're better off. It was going to be a lot of work."

I hung up, stunned by the news. We estimated that the boat would need six months on the hard, the term for when a boat is out of the water, before we could even sail her. But someone else had swooped in, taken a cursory look, and sailed away—in *our* boat. I hung up the phone, checked on the children to see that they were still playing nicely and that Sam was still napping, then stepped into the shower, where I could cry in peace. I was bereft. Where there had been reluctance, there was now regret. We had been offered a chance, and we chickened out. Would another boat come along, or had our dream just sailed away too?

That afternoon, we packed the kids into the minivan and headed down to St. Petersburg for the annual sailboat show. It was exactly the consolation we needed. Eli, Aaron, and Sarah crawled all over the new boats at the dock and tried out a little sailing dinghy (in the grass). Jay wandered through the tents looking at gear. I pushed the stroller past the authors' tents and looked at books and talked to people who had actually been out there. When the troops grew restless, I headed to the grassy field near the exit while Jay wandered through the last of the vendors' tents.

Some little girls were kicking a ball around. Our kids joined in, and then I wandered over to introduce myself to the girls' parents. Sam was napping in my baby carrier and I went to stand in the shade of the Kids Aboard tent. It turns out that Curtis and Lupe were running an educational program at boat shows, a tent where parents could drop off their kids to participate in a boat-building workshop while they shopped or went to seminars. They were homeschooling four kids of similar ages to ours, and they lived on their French-built aluminum catamaran, *Fellowship*. We were astounded to find someone who had done what we only hoped was possible. They were only in town for a few days, so I invited them to join us in Clearwater for dinner when the boat show closed. They accepted.

This was the first sailing family we had ever met. We discovered we had much in common, and they became a huge encouragement to us. Though we had just met them, they already felt like family—and we grasped for the first time how strong the bond among sailors can be. So often, the right people have crossed our path, either to encourage or be encouraged, at the exact time when it was most needed. Anyone who has ever attempted something unusual or risky knows how having someone else's support and encouragement forms an important counterpoint to the opinions of others who think their dreams are foolish, irresponsible, crazy, or all three!

Losing *Katie Rose* had taught us that regret feels worse than fear, and meeting Curtis and Lupe showed us what was possible with a large family with young children. Our kids were growing fast, and we didn't want to miss the chance to travel with them while we worked through a slow, multistep process. No matter how much practice we had on small boats in inland waterways, no matter how many baby steps we took, there would still come the unavoidable moment when we would need to actually *buy a blue water boat* in order to take our family out sailing on the unforgiving ocean. Regardless of preparation, that would still feel like a sink-or-swim moment. We wouldn't know whether we were ready, or capable, or even whether we would enjoy it, until we tried. And *not* trying would mean always wondering, and possibly regretting. With crystalline determination, we decided to skip the baby steps and take a giant leap. We began looking for a liveaboard boat, opening up the search to catamarans.

Though we knew a multihull might be more expensive than a monohull of similar length, it would provide enough space for our growing family and maybe even satisfy my husband's desire to go fast.

That is why I got so excited when, several months later, Jay came home from a weeklong business trip and announced, "I think I found the boat." Not *a* boat, but *the* boat. He opened his computer to show me and, at least on the listing, *Take Two* looked perfect. She had four cabins with double beds, a spacious interior, an enclosed cockpit (safer with small children), inboard engines with propellor shafts, a generator and watermaker, and beautiful lines. She ticked so many of our boxes and looked so attractive that we felt we should pursue the next steps, ready or not. We knew the cost of hesitation and didn't want to end up like so many other planners, armchair sailors, and readers of vicarious adventures.

All the same, we were naturally a little nervous about buying a large, custom built, wooden catamaran, built in Europe in the nineties, which had sat unwanted in Florida for three years, sustained some damage in a hurricane, and cost more than our first house. It was a risk against which we could not measure the benefits. What if it cost too much to fix? What if we hated living on the boat? What if we changed our minds and then couldn't sell it? But then, what if it was wonderful? What if this boat was the answer to our hopes and prayers? What if this was a second chance at adventure—the boat's name was *Take Two*, after all!

We reminded ourselves that we weren't committing to anything yet. We didn't have to *buy* a boat, just go *look* at a boat. Anyone can get in their car and drive to Fort Lauderdale—it doesn't take much courage to do that! We buckled our four little people into their car seats and drove our van across Alligator Alley between Naples and Fort Lauderdale. We invited Jay's dad and stepmom, Al and Mary, who were also boat owners, to come along and give us their opinion.

Looking at *Take Two* was like falling in love—sometimes you just know it's the right one. All the things that had scared other buyers away excited us. She was custom built, cold-molded marine plywood and epoxy. The designer, had thought of everything; lots of built-in storage, a roomy galley in the main cabin, an enclosed cockpit with lots of seating. But, unlike a name brand production boat, there would be no manual, no warrantees, and no company support. Parts would have to be special ordered or manufactured. The electrical system was European, fifty Hertz, and needed upgrading. In fact, everything needed upgrading—it would be a labor of love, but if we took on the project, it was an opportunity to make the boat our own. Her bones were good, her lines sleek, and the space was perfect for a large family. It was as if she had been built just for us.

While Jay was lifting every hatch and discussing systems with Al and Mary, I was corralling three excited kids while holding a baby. To them, the boat was just a new playground. After a while, I gathered them up and climbed off the boat so the adults could get down to business. At last, Jay climbed off the boat and I asked if he could watch the kids so I could have some time to look around all by myself. The next twenty minutes would likely change my life. I walked slowly around the boat, imagining what it would look like if we lived there. I climbed up into what could someday be our bunk and just lay still for a while. Even at the dock, I could feel the boat swaying beneath me. *Does a person get tired of moving all the time?* I wondered. I couldn't answer that question, nor any of a dozen others. Soon it was time to get into the car and drive back to Clearwater. It turns out that *just looking* can be dangerous; we found ourselves buzzing with excitement on the drive home as we contemplated the next risky step.

There was unanimous agreement about, and enthusiasm for, *Take Two*. She would need an out-of-the-water survey and a sea trial to tell us if she was sound, but we knew we liked her, and we felt that we could be happy living aboard. She was spacious, without the space being wasteful; she looked fast, but comfortable; and she came equipped to sail across oceans, a real blue water boat. To be fair, the kids didn't really know what they were signing up for—they thought the boat was a new jungle gym, and they imagined that they were heading out to sea like a band of pirates. What little kid wouldn't be enthusiastic? Jay's parents approved too.

But there were risks—the boat would be hard to sell if we changed our minds. There would be no turning back. Added to the fear of the unknown, there were the known fears, like bad weather, endless repairs, and seasickness. Before we could even buy the boat, there would be questions to answer. *Could we afford not only to buy it, but to fix it? Would it mean selling the house first? If not, could we secure a loan to buy the boat while we transitioned out of the house? Would we be able to insure the boat? Where would we keep a boat that size? Would we have to do a lot of work on the boat to make it livable? Were we ready to move aboard with four little kids? If we approached each step in slow motion, could we keep ourselves from feeling overwhelmed?*

The baby-step philosophy that we had discarded when we decided to skip to the end and buy a big boat would become a mantra as we made the transition from landlubbers to liveaboard sailors. We tackled the questions one at a time. First, we realized that we were not ready to sell our house and move aboard immediately, so we baby stepped our way to the bank and the insurance company. We were eligible for a loan, which we hoped to pay down when the house sold, and for insurance through Lloyd's of London, assuming we could provide a letter from a licensed captain saying that we were competent to operate the boat.

We then drove south to Bradenton to scope out a marina that had space for us. The best prospect was a side tie, which was wide enough for our catamaran, at the end of a dock at Twin Dolphin Marina. In early January 2008, we went to talk to the dockmaster, have lunch at the marina restaurant, and look at the dock space. We walked down the dock and felt the stares of the other boaters as our four little people invaded what felt like a fifty-five-and-older gated community, a common phenomenon in Florida. The marina had a pool, was located across the street from the public library and a museum with a planetarium, and was two blocks from the nice little downtown area with restaurants and shops and even a weekend farmer's market. It was up the Manatee River inside Tampa Bay, which seemed less exposed to weather and storms. And it was only one hour's drive from our home in Clearwater—far enough to be out of our neighborhood and comfort zone, but not so far away that we couldn't escape back to our house if things weren't going well on the boat. We had come to the right place.

Finally, Jay drove to Fort Lauderdale to meet the owner of *Take Two*, who had flown in from California to complete the sea trial. They would look over every part of the boat, haul it out of the water for a survey, and take it out for a sail. After that, barring any nasty surprises, we would negotiate a deal and become catamaran owners. The survey revealed items that would require a haulout to repair things that the owner called ongoing maintenance issues, but which we found troubling. During the sea trial, with the owner at the helm, *Take Two* ran aground on a shoal in a confusing channel in a Fort Lauderdale waterway. Though the owner wasn't worried about it, he promised to have a diver go down to look for damage once the boat returned to the slip. We had a decision to make.

We asked the owner to make the necessary repairs or lower the price of the boat. He refused, so we thought maybe this was a sign that we were not supposed to buy this boat. But then the path was cleared: the pictures of the hull taken by the diver after the sea trial showed damage to one of the mahogany keels. The owner was responsible for the damage and would haul the boat out of the water to make repairs. Since the boat would be out of the water anyway, he offered to address the other problems we had identified. And the deal was back on. We signed a contract, fixing a date—the point of no return.

That year, Jay was commuting out of town four days each week. I was at home in Florida with four kids under six while he worked in Pennsylvania. This was a short-term-loss/long-term-gain arrangement, where the extra money made by traveling would cover the down payment on the boat. But it left me lots of time alone after the children had gone to bed to think things over. I waffled—alternately enthusiastic and terrified about the plan to buy the boat. I remember calling Jay one night and telling him I had changed my mind. I loved our house, our neighborhood, our furniture, our friends, our safe little life. Why would I risk all that to venture out into the blue? I could feel his frustration across the miles. "Woman! I need your support!"

Jay says the biggest contributing factor to success on a boat is a happy woman. It's not the kind of boat, the weather, or the travel plans, but rather the mental and emotional state of the first mate. He's obviously done a good job of managing my emotions because we would never have even made the purchase without his steady voice of reason. He thought I was just having a case of cold feet and told me to sleep on it. I called a girlfriend and she talked me off the ledge, reminding me of all the reasons I had wanted to pursue this dream in the first place. The next morning, I felt better, and after that fit of anxiety passed, we moved forward without reservation.

Just before Jay's thirty-third birthday, we signed papers and became the proud but trepidatious owners of a forty-eight-foot catamaran that was built while we were still in high school. It was wider than *Blue Bear* was long! (Our sweet little day sailer would soon be sold because we couldn't focus on more than one floating thing at a time.) That April, we drove to Fort Lauderdale again, this time to clean the boat and get her ready for the delivery.

We dropped the three older kids at my brother's house in Naples to sleep over with their cousins, and took Sam, who was a year old and still nursing, with us to the boatyard where *Take Two* sat on the hard. Her keel and decks were repaired, and she was almost ready for launch. We stayed a few nights in a little motel nearby and handed Sam up the ladder every morning so we could begin working on a project list while he napped on a settee. We dug through lockers and storage compartments, sorted tools and gear, made sure the basic systems worked, and cleaned up the living space for Jay, his dad, and the delivery captain we had hired to bring the boat from Fort Lauderdale to Tampa Bay.

After fifteen years of dreaming and thinking and planning, we had finally done what we thought might not be possible: we bought a boat on which we could live and travel. All our preparation had led us to this moment, and we were about to find out whether we would sink or swim. For as long as we humans have ventured into the unknown—whether it's something as dramatic as discovering a new continent or landing on the moon, or as mundane as starting a new business or moving cross-country—we have been willing to swap the known Today for the hope of a better Tomorrow. We may be driven by curiosity, despair, optimism, glory, boredom, or the promise of freedom or financial gain, but whatever the case, the drive to explore, dream, and discover seems to be built in. For Jay and me, faith gave us the courage to dream big, and courage spurred us to defy fear and take a giant leap. Whether that leap would look like flying or falling, it was still worth a try.

# 4

## RUNNING A TIGHT SHIP
### DISCIPLINE

*May 2008. Jay sailed* Take Two *into Tampa Bay today. I'm packing an overnight bag and trying to corral the kids so we can meet him at the dock. He thought I should wait until tomorrow, but I have been waiting for this moment for more than fifteen years and I didn't even get to see the boat come to the dock. I am supposed to go to a homeschool book fair at a friend's house, but Jay and his crew have come in a day early. I call and tell my friend that I have to bail. She says, "You've waited this long, can't you wait one more day?" The answer, of course, is, "No!"*

*We drive across the Skyway Bridge and Tampa Bay glistens in the late afternoon sunlight. So many good things are about to happen—I am nearly bursting! We park at the marina and walk proudly down the dock: we can see* Take Two's *signature blue down at the end. But when we get there, Jay is tired from the trip and worried about the safety of the boat for the kids. We have a one-year-old who puts everything into his mouth and three children who bounce off the walls. They are only four, five, and six years old, so it's not surprising. To make matters worse, a forward-facing cabin window has broken during a brisk sail when the jib sheet got caught under the slightly* ajar edge and the force of the wind in the sail ripped it right out of the frame, breaking the safety glass into a thousand tiny cubes and spraying them all over the interior. We will probably be finding these little glass cubes for months.

Jay gives the kids a tour of the boat, explaining how things work—especially safety items like the fire extinguisher (don't touch it!) and the sea cock in the head, which, if opened, could flood the boat with sea water (don't touch it!), and the bright red battery switch at toddler height (don't touch it!). He also shows them the A/C electrical panel switches (don't touch them!) and the knobs for the stove (don't touch them!). The kids, who up until this moment thought that this was simply a new playground, are beginning to look doubtful. All but the toddler begin to understand that the floating house is also a vehicle, with complicated systems. It will be a big adventure, yes, but it will take time to make the boat comfortable and safe for a family. Until then, there will be a lot of "nos" and "don'ts."

We find a frozen pizza that the delivery crew hasn't eaten and put it into the oven and sit the kids down in the cockpit for dinner. I snap a quick photo of our little people having pizza as the sun sets; then Jay and I clean up inside and get sleeping spaces ready. Thus ends the first day of the rest of our lives. Tomorrow, we'll go back home and form a plan of attack for beginning our boat life. We just need to make it through one night.

⌇

Old wooden ships were said to be "tight" when the seams were well caulked and the lines were taut—everything orderly and made ready to set sail. A captain who ran such a ship kept strict discipline, adhered to a maintenance schedule, and made and enforced rules consistently. A tight ship gave the crew a sense of security—a well-maintained boat and clear expectations makes everyone safer at sea. Even now, long after the era of wooden sailing ships has come to an end, we use this phrase to describe a boss who keeps an orderly office environment and expects a lot of his employees, or of parents and teachers who have well-disciplined children. Without order and organization, a boat, like an office, a family, or a classroom, can fall into disarray and become an unpredictable or even unsafe environment, especially during times of stress. For humans, who often take the path of least resistance, nothing can be more challenging than maintaining good habits and consistency.

One of the ways we run a tight ship on *Take Two* is by posting rules and expectations. When I was a kindergarten teacher in a public school, I had access to all sorts of fun equipment, and the laminator was probably my favorite. I love laminated posters. Even now, I carry a small laminator and plastic sheets to waterproof and immortalize information. Posted signs are a visual reminder—a way to help people organize themselves and their environment. When the kids were little, for example, we posted *The Night List*, which had pictorial cues and large print so I wouldn't have to say "Did you brush your teeth?" or "Did you get a drink?" half a dozen times. When the night list was done, I would read the bedtime story and turn out the light. *The Chore Chart* still occupies a central location on our boat—and no crew member can claim ignorance of his duties. Other things, like *The Clean Bathroom Checklist* or *The Ten Commandments for the Table* (a tongue-in-cheek "thou shalt" list of table manners) are posted in the area where the reminders are necessary.

I recently took down a dog-eared poster that was once at eye level for the majority of our family but had become virtually invisible as crew members literally outgrew it. It was sun-faded and had crinkled edges, and the double-sided sticky tape that held it up was so old that it left permanent residue when I removed it.

*The Boat Rules* had been posted for so long that we had internalized the instructions:

1. *No running, jumping, or climbing.*
2. *No touching buttons, levers, or switches.*
3. *Always ask permission to go on deck.*
4. *No screaming, shouting, or whining.*
5. *Obey the captain immediately, and without question.*

*BE SAFE, HAVE FUN!*

Just imagine a boatload of kids who never whined! Who never ran overhead on deck like a herd of charging rhinos! Who never inadvertently turned off the electrical system with the flip of a switch! Who never shouted from one hull to the other! The list tells as much about our idealism as it does the stage of life in which we made the rules. In the beginning, we still had toddlers to keep alive and we lived in a marina where we feared the complaints of nearby neighbors. This is not the Royal Navy, but without a chain of command, without boundaries, children on a boat are not safe and not fun to be around. Also, when your yard is made of water, the risk of losing a life is real and terrifying.

All of our kids were well trained in basic obedience—responding to their names, coming when called, respecting "no," and generally listening to adults. They were also trained for safety—to come when we rang the bell for "all hands on deck," for example. They all learned how to swim at an early age, but we required life jackets when the boat was underway, and even at anchor for the youngest ones. They were required to ask permission to leave the cockpit so we would know where everyone was, to make sure that no one could fall in without our knowledge. We had safety protocols and flotation devices, but we relied primarily on rules and consequences. Even in a house, training a toddler requires undivided attention, vigilance, and consistency. On a boat, you can't take your eyes off of them for a second.

The best rule, of course, is all-encompassing, nonnegotiable, and in perpetuity: *obey the captain.* Immediately. Without question. (And *obey the first mate, by order of the captain!*) And how about that last little note? *Be safe, have fun!* After they had obeyed all the other rules, which would entail making no noise and moving only slowly and carefully, how could a child be expected to have fun? We realized that some of the rules were unrealistic, especially when we were at anchor. Some of the favorite activities involved climbing the rigging, jumping on the trampolines, and leaping from the aluminum arch we had installed, complete with monkey bars and a diving board. Of course, if things became unruly, we could cite the rules to get things back under control.

Rules by themselves are meaningless without consequences: when disobedience occurs, we correct with words and actions. We use logical consequences whenever we can, and when necessary, physical reminders. For older children, disrespect or loss of self-control results in pushups, extra chores, loss of privileges, or some time alone to pull themselves together. On the flip side, we have a system in place for noticing and rewarding good behavior, including a fun outing with just Mom or Dad, a rare treat in a large family.

Unfortunately, the surest way to learn a lesson is to suffer the natural consequences. On a boat, this can be physically painful. One memorable example was when the children ignored the guidance not to play with doors. During an exciting game of hide and seek, during which there was probably a fair amount of illegal running, Sam's happy toddler giggling suddenly turned to screaming. He had gotten his fingers smashed in the piano hinge of our pantry doors, resulting in the loss of a fingernail. No one needed to say, "I told you so," because everyone knew better and felt badly. Once, Sarah stepped on a deck hatch, something we had warned against over and over because we didn't want anyone to slip and fall and we didn't want the expensive hatches damaged. This time, instead of stepping *on* the hatch, she unexpectedly stepped right *through* the hatch, which someone had left wide open. The resulting fall into the port hull could have been so much worse, and she was lucky to escape with only a few bruises. Pain being a good teacher, we have all learned lessons the hard way, and the more painful the consequence, the more memorable the lesson. This may sound harsh, but we believe in letting our kids discover their own strengths and limitations, within reason.

One of the side effects of our choice to live on a boat is that our family participates in activities that some might consider unsafe. But if we had wanted safe, boring lives, we would have stayed in the suburbs. Because we allow active things like climbing rigging, jumping from waterfalls, free diving, piloting the dinghy, spearfishing, and wakeboarding, we have to trust our children to obey rules and take steps to keep themselves safe. If someone goes off in the dinghy, for example, that person carries a handheld radio and a few basic emergency supplies. They wear life jackets, an engine kill switch (so the boat won't run them over if they fall out), and return at an agreed time. We have raised them with the expectation that they will be responsible—for their behavior, their things, and their time—and they earn freedom as they show themselves trustworthy.

Aside from safety, there are other good reasons to have well-disciplined children: to raise people with good character, to give them success in the future, to make them pleasant to be around, and to be welcomed back. When we come to a new place, for example, we tighten the reins and raise expectations for our kids, so that we aren't the family with a brood of out-of-control kids running around. After we have proven ourselves, we can loosen up a little, but by then, we have demonstrated to the people around us that our kids are aware of their limits and respond when corrected. And they are expected to be polite—because *please, thank you,* and *you're welcome* are keys that open all sorts of doors.

Once, motoring toward a marina we had previously called home, we radioed to see if they had a slip available for the season. "This is sailing vessel *Take Two*. We'd like to come in on a side tie. Over."

"*Take Two*, what's your overall length and how long would you like to stay? Over."

"We're a forty-eight-foot catamaran, and we'd like to stay the whole season, if there's availability. Over."

"I'm sorry, we're all booked up. All we have is two nights on the transient dock. Radio when you're ready to come in and we'll have a dockhand there to help you tie up. Over."

"Thanks. *Take Two* standing by on channel sixteen." I turned to Jay. "Did you hear that? What are we going to do now? Should we keep going and find another place to stop?"

"We'll figure something out. Don't worry about it. We'll go in for the night and think about it tomorrow."

Twenty minutes later, the dockmaster radioed back. I recognized the familiar and friendly voice immediately.

"Hi, Tanya! This is Dan. I'm so sorry I wasn't in the office when you called! That was the new guy and he doesn't know you guys. I just want you to know that you're welcome to come in and stay as long as you like. We can always find space for *Take Two.*"

I thanked him and hung up the radio handset. I then did a

little happy dance and gloated: "See how all that good behavior pays off?!"

The rules and consequences not only established limits and safe practices and made our kids more enjoyable, but they also addressed a deeper need for order. Maintaining a boat and a household of seven is so chaotic in and of itself that without regular routines and clear expectations, we would become quickly overwhelmed. Beyond the dictionary meaning of discipline, *the practice of training people to obey rules or a code of behavior, using punishment to correct disobedience,* there is a connotation of basic respect, consistency, and fairness that creates a secure and predictable environment despite changing surroundings. And we taught our kids that the end goal of discipline is *self-discipline*: if you govern yourself, you won't need to be governed.

This kind of discipline is the hardest to master: to make yourself do what you don't want to do, to control your own negative responses, and to be consistent in your good habits. The word "discipline" is based on the Latin root meaning "instruction" (just as a "disciple" is one who learns from a teacher), and the boat has been our classroom. Without external controls (a time clock at work, a school schedule, or rush hour traffic), self-discipline is critical to making our life work—to earning money, to maintaining our boat, to eating well, to homeschooling our children, and to making travel plans. Living on a boat has taught us how important it is to order our thoughts, our routines, and our environment.

And here is where I make my confession: I am terrible at self-discipline. I wasn't good at it even when I did have calendars, set appointments, and a work schedule. I have no sense of clock time, so I am often late. I am not, by nature, a patient person. I love to make plans but struggle to complete projects. I have not only a *childlike* exuberance but also a *childish* moodiness. I worry and fret. I lose things, forget things, and am easily distracted. I talk more than I listen. I literally have *all* the symptoms of ADHD in adults. I have spent my life trying to contain my inner chaos: making project lists, writing down personal goals, creating daily and weekly routines, coming up with meal plans and weekly menus, organizing homeschool plans and schedules, labeling bins and sorting our physical stuff, and joining fitness classes to make myself exercise. I have a morning routine that starts with prayer and a devotional reading to encourage peace and mindfulness. These coping mechanisms helped a lot when we moved aboard; too much freedom can be dangerous to people like me!

Thankfully, I'm married to someone who has more than his fair share of self-discipline. He sends his thoughts through a filter before speaking his mind (imagine that!), rarely gets angry, and works really hard, even when there's no boss telling him what to do. When Jay left his brick-and-mortar job in Atlanta, he began building skills and a clientele that would allow him to work remotely. He once adhered to a dress code of button-down shirts, wool slacks, and loafers, but now he commutes to his consulting job barefoot in a pair of shorts, walking from the starboard hull to the port hull, where he works in a tiny home office using high-speed internet to connect with clients and work on projects in cities far away. He makes himself sit or stand in front of his computer for as long as is necessary to complete a task, solve a problem, and pay the bills. There are always competing interests—boat projects to complete, opportunities to snorkel or spearfish, or fun field trips to experience with the family. Self-employment means that no one tells him how much to work or play; he has to figure it out for himself. My spontaneity and desire for change help shake things up, but without Jay's stability and dedication to hard work, we would not be able to live this way. He helps set an example for our children, as well.

For us, and for many families we meet, the hardest aspect of moving aboard a boat is the full-time job of educating the children. Home education has become more popular since I was a kid, resulting in the availability of countless methods and curriculum choices. I went to the Florida Parent Educator's Association Homeschool Convention the year before we bought the boat to buy materials and get ideas for the first few years of elementary school. It was overwhelming, to say the least, but as I waded through the vendor's tables and sat through workshops led by veteran home educators, I began to realize that I could tailor our homeschool to fit our lifestyle and to meet the individual needs of my students. The real challenge, we would come to realize, isn't educational, but relational.

I used to teach in a kindergarten classroom where the day started with the Pledge of Allegiance and ended with the ring of a bell and a rush of departing students. I was paid to teach other people's children, and I felt loved and valued every day. Now I manage a floating one-room schoolhouse—for intrinsic rewards only—with kids anywhere from preschool age to college age, in the same environment where we also *live*. All the demands of keeping a home tidy, planning meals, traveling, writing, and having fun coexist and efforts are often taken for granted. Personal problems and distractions don't go away at the ring of a bell. The students are sometimes cranky, resistant, or unfocused. The teacher is sometimes cranky, resistant, or unfocused. We often do better when we have places to go, people to meet, and a routine to work around, but at the same time, we don't live under the tyranny of a clock and calendar.

While our children never went to a formal school or participated in team sports that had practice schedules, we always stressed work before play. School days took on a predictable order: wake up, have breakfast, listen to a read aloud for history, literature, or science, and then get started on the independent work like math practice, spelling, or journal entries, with the teacher-mom bouncing between students as they had needs and questions. Sometimes Jay gets involved with special projects, but for the most part, he fills the role of principal, supporting the teacher and enforcing guidelines when necessary. We usually stop for lunch and finish formal schooling in the early afternoon. When the daily requirements are met, the kids are free to explore their own interests, which equates to learning for pleasure, like practicing a musical instrument, fishing, drawing, and working on LEGO projects. In addition, there have been lots of spontaneous learning opportunities during the day (a.k.a. interruptions), like talking to a neighbor who stops by in their dinghy, watching dolphins splash around the boat, baking bread, or hanging laundry on the lifelines. We have taken a lot of field trips and integrated our environment into our learning. Marine biology and the habits of sea birds, for example, dominated our science curriculum for the first few years.

Each of us has had our triumphs and struggles with homeschooling. Learning is done for the purpose of mastering a new skill, not for getting a grade, but, as a result, deadlines are arbitrary, and getting a student to finish a writing assignment, for example, is a challenge, as is getting the teacher to correct it in a timely manner. Schedules are relaxed, and the line between work and play blurred, making it hard to stay focused. Ideally, boat life offers us the chance to be disciplined enough to accomplish important tasks but not so regimented that we suffer the negative effects of stress and fear of failure. Not that we don't fail, but rather that we view the failures as opportunities to do better next time.

Whether we're talking about work, home management, or school, we are all still learning this important life skill: how to balance *have-tos* and *want-tos* as we govern ourselves. We are often walking a fine line—our boat is not falling apart, but there are always unfinished projects. Our family is not dysfunctional, but there are always ongoing challenges. There are tasks that must be completed to keep our home running, but how we divide that work is a topic of lively discussion. Every household, floating or not, must decide how tightly to run the ship: how to find the equilibrium between obligations and leisure time, between rule making and rule breaking, and between safety and adventure.

I would love to tell you that we always run a tight ship, that living on the boat has taught us how to lead disciplined and productive lives outside the rules that govern so-called normal society. I wish that we had learned all our hard lessons the first time around. But the truth is more compelling: we are ordinary people who are attempting something extraordinary and sometimes succeeding.

# 5

## LEARNING THE ROPES
### MAKING MISTAKES

*June 2008.* We feel like we have stepped into a fairytale. Music drifts over the water, pretty lights twinkle in the trees on shore, and a breeze is blowing, rocking us gently and making a distinctive whirring sound in the rigging of the boats on the dock. We are spending the weekend on the boat, one of our first, and we are still incredulous that this is really happening. All the sacrifices, all the hard work, all the years of hoping and praying have come to fruition. That we have not actually sailed our boat doesn't bother us—we finally have the means to do so, and for tonight, that is enough.

We have just tucked the kids into their cozy cabins, finished hand washing the dishes, and opened a bottle of wine. We are getting ready to take our drinks out to the foredeck when I smell something. I sniff around and ask Jay if he smells it, like something hot, or electrical, like melting plastic or maybe even burning wires? Yes, he smells something, but can't tell where it's coming from. We don't see any smoke or find any obvious signs of a fire. We sniff around, trying to locate the source of the smell. All the systems on the boat are new to us, and we are using things for the first time. It smells like it's coming from near the refrigerator or freezer, but those things seem to be functioning without a problem. We rummage around inside lockers and look under and around things but cannot find anything wrong. We look at each other and shrug. We pick up our sturdy acrylic wine glasses and head outside.

After a pleasant evening, we come back in. The mysterious smell has disappeared, and we head to bed. The next morning, I mix up pancake batter and pull out some bacon for Sunday breakfast. I turn on the ancient Bosch electric cooktop, but nothing happens. "Hmmmm…" I wonder aloud, "there seems to be an electrical problem…could it be connected to that mysterious burning smell from last night?"

Jay begins to earnestly chase down the problem, following old wiring from one end of the boat to the other. I decide to take a chance on breakfast and do something I would never do if we lived in a house: I go outside and talk to the neighbors, who are enjoying coffee on the dock. I explain the problem and ask if they know how I could turn my bowl of batter into pancakes for hungry kids. My neighbor, Joanne, does not tell me to go get doughnuts instead, or even offer me an electric griddle to plug in on the dock—she simply invites me into her boat to cook pancakes on her stove. I am blown away by her generosity and offer to share the final product with her and Pete.

Meanwhile, Jay finds a wire that was slightly corroded and had too much current running through it. It overheated, melted the outer coating of the wire, shorted out, and burned the transfer switch. Turns out the ignore-it-and-it-will-go-away method of problem solving doesn't really work on a boat. Thankfully, our mistake didn't lead to an electrical fire. A boat can burn to the waterline in ten minutes, giving the crew very little time to escape. We draw several conclusions: we cannot afford to ignore suspicious sounds or smells, we should check every wire on Take Two, and we need a fire escape plan to practice with our children. Also, boaters make the best neighbors.

The first time we brought *Take Two* to the dock, the momentum of our excitement ran right into the learning curve, drove up its steep side, and sputtered to a crawl. Though Jay had grown up sailing, and I'd had a few experiences in small boats, we had never been on anything this *big* before. Everything on the boat seemed big—the mast is sixty-eight feet off the water, the anchor chain is a hundred and fifty feet long, the winches are like drums that can handle lines with a diameter of three quarters of an inch, the main sail weighs hundreds of pounds, and the zipper on the stack pack, which covers the sail as it lays on the boom, is nineteen feet long. I know because I repaired it.

I remember learning how to "reef" or "shorten sail," an important skill for reducing the area of the mainsail in heavy weather. It's something you want to practice before the need arises and entails lowering the triangle of mainsail attached to the mast so that the size of the cloth is reduced (thus diminishing windage and boat speed). I can pinpoint in memory the windless day Jay showed me our reefing system, and how we practiced raising and lowering the main while safely tied to the dock. We have three reefing points on our main, and each point has a line for attaching the bottom, or foot of the sail, to the boom. Each line is a different color and runs from an outhaul winch or a cleat on the mast, through a sheave on the boom, down inside to another sheave at the other end, where it is attached to a grommet near the leach of the sail and back down to the boom, where it can become the new clew of a reduced sail.

If this seems complicated, that's because it is, at least the first few times, and especially when we are sailing, watching a squall approach from a black horizon, whipping the sea into a froth. One person is at the helm, keeping the boat pointed directly upwind so the other person can go forward to the mast to shorten sail, creating a smaller triangle of cloth. (Technically, it involves lowering the halyard, creating a new tack by attaching a line from one side of the mast through a cringle on the lowered sail to the other side of the mast, taking up slack on the halyard to retention the luff, and tightening and locking off the correct reefing line—which takes almost as much effort to describe as to accomplish!) The phrase *learning the ropes* conjures the memory of that day, my feeling of ineptitude as I tried to remember what line connected to which part of the sail and telling myself *not to screw up*.

Learning how to sail, how to homeschool, how to use epoxy to fix wood rot, how to make bread—all of these skills pale in comparison to one of the most important lessons we have learned on our boat: how to make a mistake. Jay and I both share the coveted position of oldest child in our families, something that psychologists say affects our overall approach to life. We are both goal driven, performance-oriented, and hard on ourselves. I know this pressure rolls downhill, and despite our best efforts to have a relaxed attitude, our kids fear making mistakes. But nothing is more human than making a mistake—and nothing offers us a better opportunity to learn. And so, despite how we *feel* about them, after our initial reactions, this is the logical protocol we follow when we make a mistake:

*Admit it.*
*Apologize for it.*
*Fix it.*
*Learn from it.*

One learns the ropes quickly on a boat because mistakes can be costly, both in dollars and bodily harm. A misstep can cause a broken toe, forgetting to duck can result in a knock to the noggin, failing to tie a line properly can mean a lost dinghy, and neglecting basic maintenance can literally sink a boat.

Opportunities to learn are plentiful, starting from the moment one first steps onto the boat at the dock. Quick learners do not have to repeat the same lesson twice, though there always seems to be something new to learn. These "lessons" fall into general categories on a boat: injury and accident; trial and error; breakage, leaks, and fire; mishandling; and poor planning. None of the mistakes signify mere nautical errors, however; all of them carry life lessons.

At the beginning, it seemed like things broke faster than we could fix them. Weekends were spent making basic systems operational and cleaning and organizing the living space so we would be more comfortable. In addition, we were learning about the physical limitations of our new, smaller space. I hit my head going down the companionway stairs to my cabin more times than I would like to admit. The children stubbed toes on bulkheads, slipped on wet hatches they had been told not to step on, fell in the water while goofing off on the dock, pinched fingers, bashed shins, and used so many Band-Aids that I wished I had bought stock in Johnson & Johnson. While cleaning and working on the boat, we got splinters, burned our hands, blistered and bled, then retreated bruised and battered to our comparatively comfortable house to heal in time for the next weekend aboard. Without determination and a rapid toughening up, we might have thrown in the towel that first experimental year.

In addition to physical challenges, there were mental and emotional obstacles to be surmounted. Learning something new can bring joy as well as frustration, and can test the roles in a relationship. Few things are more frustrating on a boat than docking and anchoring. Whole books have been written about the subject—and Jay has read them. I went to a women's sailing seminar and took two courses, one in basic sailing and another in docking and anchoring, but when it came time to pull away from the dock, we broke out in a cold sweat. Nothing like motoring past million dollar yachts in a two-knot current with wind on the beam to elevate the heart rate. Accordingly, our insurance company had insisted that we have written proof that we were capable of handling our boat, so we did a few docking drills with the delivery captain before he would sign our paperwork. It turns out that docking and anchoring is not so much about moving a large boat in a small space with the forces of wind and water acting upon it as it is about good communication between captain and crew, including a good system of nonverbal signals.

In order to better understand what was required of the helmsman and the crew on the bow, we also tried a little role reversal. Redundancy is a familiar concept on a catamaran, where there are often two of everything, so having both people proficient in both jobs seemed like a logical approach. I spent some time piloting in the river outside the marina, circling the boat and steering between two buoys that Jay had anchored, or practicing the equivalent of parallel parking against them, skills needed for docking. When I was ready, I brought the boat into the dock for the first time, Jay coaching me as I gently throttled the twin engines to lay the boat aside our pilings. I maneuvered flawlessly into our slip. However, while Jay was helping me to do his job, no one was doing my job, which was to put out fenders, get dock lines ready, and make sure we didn't hit anything on the way in. A crunching sound from the stern on the starboard side told us what we had forgotten. Despite the benefit of redundancy, we also learned to recognize that sometimes one person is better at something (like Jay mixing and applying epoxy to repair damage), and we have a new appreciation for our complementary skill set.

I don't ask Jay to cook meals or homeschool the kids. I may

ask for menu suggestions or show him a piece of our curriculum, but these things are not his forte. And he doesn't ask me to fix things or design a database. He has an engineering brain: when there's a problem, he solves it. Some things I have learned to troubleshoot, and he's coached me through propane tank changes and watermaker procedures, but I don't enjoy it. To each his—or her—own, I say.

After six months on the dock, learning, repairing, cleaning, and organizing, we were ready for our first cruise on *Take Two*. We had been sailing in Tampa Bay a few times and had anchored overnight in nearby bays. We had learned to use the generator and watermaker to make power and water when away from the dock. We had accrued all the required safety equipment and studied the charts for nearby coastal waters. We had a new dinghy, a Porta Bote lightweight, folding boat with a Yamaha 8 hp (eight horsepower) motor. We had even chosen a departure date and a destination: the week of Thanksgiving, we would sail south to Charlotte Harbor to meet with Jay's parents, Al and Mary, on their boat, *Double Entry*, where we would anchor together near Cayo Costa State Park and share turkey and trimmings.

In order to get a head start on the travel day, we anchored the night before the passage at the mouth of the Manatee River. Immediately there were problems. The battery bank wasn't charging properly. The watermaker malfunctioned. And the toilet wouldn't flush. These three systems—power, water, and waste—are all critical for a week on the water with six people aboard. Jay kicked it into high gear, discovered that we could operate the boat on one battery bank (yay, redundancy!) and run the generator twice as often. He fixed the watermaker so we wouldn't have to ration water (yay, showers!) or cut our trip short. And he fixed the toilet…but not in time for the morning rush. So we pulled out the emergency toilet seat that fits on a five-gallon pail and used the bucket-and-chuck-it method. It was not exactly what we signed up for but worked in a pinch.

On the way out of Tampa Bay, hoping to cut some time off the passage, we followed the chart toward Passage Key, a shortcut through a shallow sandy area that doubled as a nudist beach on the weekends. On that blustery November day, there was no sign of a sandbar, only wind and waves and a hidden narrow cut. We were under full sail, going about eight knots (faster than it sounds), and trying to find the entrance to the unmarked channel. Jay looked at me and said, "What do you think? Should we thread the needle?"

I stammered something unintelligent like, "Uh…you're the captain. What do you think?"

"Do you see those breakers?" he asked. "There's a place in the middle where it looks calm. But it doesn't line up on the chart."

"Yeah…maybe we should stay away from the breakers… What's the risk here? Like if we run aground?"

"At eight knots? Under full sail? Maybe a dismasting."

That didn't sound good. Was it worth risking the boat to arrive in port a couple hours early? How was I supposed to know? I'm just the first mate and galley wench. I deferred to the captain. If we stood there discussing it any longer, the choice would be made for us.

"Do whatever you think is best. If it looks okay, go for it." So he did.

I went inside to stand at the galley sink and wash dishes to keep my mind off the danger. I was humming happy little songs to myself and taking deep breaths. We bumped the bottom. This is not like running over the curb in your car. It's more like setting your house down on a witch after traveling by tornado. All the dishes rattled and the rig gave a shudder. *We're not going to make it*, I thought. The waves picked us up, then we bumped again. I closed my eyes and said a wee prayer. Then we bounced past the shallow spot. I opened my eyes. *We made it.* We passed narrowly between the areas of rough water and out into the Gulf. Afterwards, as we pointed the bow south and headed towards our destination, we debriefed and made a firm decision: it is not worth risking the boat to shave a couple hours off of a passage.

Without further mishap, we sailed down the coast and stopped for the night at an anchorage near Venice. The next afternoon we arrived at our destination just inside the Boca Grande pass, Pelican Bay. We anchored near Jay's parents and the children asked if they could swim over to visit Grandma Mary and Skipper. While they put on their wetsuits to jump in the chilly water, Jay opened a beer and I made a gin and tonic. This called for a celebration.

The next day, we unfolded our Porta Bote, attached the motor, and lowered the dinghy into the water using a halyard and a winch. We took a test drive and then planned our lunch outing for Cabbage Key, a nearby island resort with cottages and a restaurant that can only be reached by boat. The week was filled with similar delights, which felt like recompense for the six months of work we had done on the boat. Jay and his dad took Eli, Aaron, and Sarah out fishing while Mary and I visited over coffee in our cockpit during Sam's nap. One day, we motored over and tied the dinghy to the dock at Cayo Costa's ranger station and clambered ashore to stretch our legs on the endless white sand Gulf beach. We walked through hardwood hammocks and coastal forest preserves, gorgeous views of a Florida that is elsewhere mostly buried under beach condos and strip malls. Eli, our seven-year-old nature-loving boy, collected millipedes and built a habitat for them in a plexiglass aquarium we had given him for storing temporary pets. Aaron, an active six-year-old, played on the beach, trailblazed through the forest, and swam. Sarah, our blonde-haired five-year-old, collected shells and built sandcastles. Sam, at age two, was obsessed with seabirds and loved chasing seagulls, pointing out "pecalins," and watching sandpipers run along the beach. We had a Thanksgiving meal on *Take Two*, complete with pumpkin pie baked in a galley oven. We were especially grateful that year: we had gotten a taste of what cruising was like—a large success that balanced out many small failures.

At the end of the week, we checked the weather forecast carefully and planned our trip back to Tampa Bay. We waved goodbye as Al and Mary turned south toward Naples and we headed north in the Gulf of Mexico. We quickly realized that neither the weather nor the sea state was exactly what we had expected, something we have since come to accept as a common occurrence. The wind had died but left behind six-foot swells. We had to motor north with seas on the beam, rocking and rolling us for the better part of a day. Since we are a catamaran and don't heel over, we are somewhat lackadaisical about stowing loose items when sailing. But when a particularly large wave hit our starboard side that evening and rang the ship's bell, I heard a crash-bang-splash and knew we had trouble.

In preparation for a long day of sailing, I had made a pot my famous Thanksgiving Leftovers Soup with Stuffing-Dumplings, which I had warmed up for lunch that chilly November afternoon. Despite lackluster appetites, we had each had a mug of soup. Doubtless, I should have put the leftovers in a secure location, but with the uncomfortable motion, I had left the covered pot on a trivet on the counter and decided that cleanup could wait until we were in calmer waters. Indeed, it could not. I ended up mopping the greasy leftovers off the galley floor and cabinetry as best I could and tucking a little note to self in the corner of my mind: *take the extra time to put things away*. Both Jay and I learned the ropes on that first cruise. On a boat, as in life, laziness is often rewarded with more work, not less. As J.R.R. Tolkien said, "shortcuts make long delays."

From these and many other mistakes we made that first year, we began to forge a new lifestyle and life philosophy: failure is a necessary part of learning. We screwed up more often than we succeeded, we held debriefings, made apologies, and offered forgiveness. Sometimes when things went wrong, we were prepared, and other times we were lucky. When we were neither, we made amends and vowed to do better. We learned about sailing in summer thunderstorms, setting the anchor, properly securing the dinghy, and closing the hatches at night in case of a sudden rainstorm. I can't speak for the rest of the crew, but I often felt frustrated and discouraged, and more than once I wondered if we had taken on more than we could handle.

We met another sailing family on the dock at Twin Dolphin Marina who taught us an important lesson about making mistakes. Dave and Vicki lived on a fifty-three-foot monohull, *Odyssea*, with their four boys. They were the only other people with children on a boat in our area, and we were instantly drawn to them. In time, they became like family. Vicki was watching our kids one morning while I ran an errand. Our firstborn perfectionist, Eli, was helping make French toast. He stood by with the bag of confectioner's sugar for sprinkling the plate of warm bread, but a group of kids moving in and out of a boat galley resembles a pinball game more than a cooking class. Somehow the bag got knocked out of his hand and the contents exploded onto the floor and sent a cloud of powder into the air, which then sifted down like a fine, white snow. My poor shell-shocked child immediately began apologizing, but Vicki, instead of getting upset, began to laugh hysterically. They cleaned up the mess and no more was said about it.

But the kids told me all about it that afternoon and I was impressed by Vicki's response. It struck me that there was a missing step in our mistake protocol: *laugh about it*. We are entirely too serious for our own good. Sure, we laugh about our mistakes *afterward*; they make the best stories. But if the boat represents a school of daily life lessons, then laughing about blunders and bloopers is a marked improvement over yelling, swearing, criticizing, moping, or regretting. Learning the ropes requires humility and patience—and the ability to laugh at ourselves and move on. Whatever our attitude toward them at the time, we are ultimately grateful for every error, miscalculation, oversight, mishap, and faux pas we make: they pave the path to progress and success. They make us who we are.

# 6

## CLOSE QUARTERS
### CONFLICT RESOLUTION

*October 2008. It's a classic situation: a boat comes into the anchorage, woman at the bow ready to deploy the ground tackle, man at the wheel, shouting instructions. Conflict arises, they have an argument, and thanks to the way in which sound travels over water, they provide free entertainment for the whole harbor.* We know this, we have witnessed it firsthand, and yet when a misunderstanding arises one weekend while learning to anchor, I find myself on the bow with the windlass control in my hand, shouting back at Jay, who can barely hear me, getting more and more frustrated. Add to that picture four noisy kids creating chaos in the cabin, and Lord only knows what the neighbors think.

We finally get the hook down and bridles tied, and when the boat falls back, we turn the engines off. Jay steps out of the cockpit, and we debrief. I'm unhappy about how it went but not sure what went wrong. He lets me know, but I feel criticized and try to explain what he couldn't see from his place at the helm. I explain loudly and with emotion. He says that if I want to stand outside and yell, that's my business, but he's going in for a beer.

What can I do? I stand there feeling angry, but there's no one left *to argue with and the kids come out to play on deck now that the anchor is down. It takes me a few minutes to cool down and decide it's not worth it to keep feeling upset. I realize I still have a lot to learn and that my childish emotional response didn't help. I get a drink, find Jay, and apologize—not for the first time, and not for the last.*

*Because this situation, though the exact circumstances may change, has happened before and will happen again. Nowhere have we so often been forced to face character flaws and personality differences as in this small space from which there is no escape. The boat has become a crucible in which all the scum rises to the surface. If we can skim it off, then next time the heat is turned up, in theory at least, there will be a little less trouble and more pure enjoyment.*

Our goal to move a family of six who lived in a house with two-thousand square feet of living space and a large yard to a forty-eight-by-twenty-six-foot boat tied to a 100-foot dock was ambitious, to say the least. With only four cabins, each person possessing his own private space would not be an option. Until we replumbed the heads, we would have to share a single, hand pump, saltwater toilet. To say this was *close quarters* is not an exaggeration. Our catamaran is roomy and comfortable, with plenty of seating in the salon and cockpit, and the foredeck is open and spacious with nets (called trampolines) between the hulls for lounging on. To boat people, our catamaran seems huge. To house people, on the other hand, it feels cramped. There is nowhere you can go to be truly alone—we are intimately aware of every other person's whereabouts and we are never out of earshot of each other. Sometimes proximity equates to intimacy, and other times it just feels claustrophobic, like a burrow full of meerkats. Conflict is inevitable and privacy is impossible. Anyone who moves onto a boat must deal with these *shrinking pains*.

Many fellow liveaboards are young couples out cruising before or instead of settling down to have kids, or retired couples out traveling after their kids are grown. They must learn to coexist in a small space where the only constant is change. Often, it's a man with a dream who's found a woman, or convinced one, to go along with his boyish idea of a seafaring adventure. We have met many single men (and a few single women) who started out with a partner but found the pressures on the relationship to be too great to continue together. The options when things go south are somewhat limited: learn to put up with each other, sell the boat and go back to the normal life (where you may or may not stay together), or split up—one person jumps ship while the other continues the journey alone. Sometimes we meet solo sailors who opted to go alone, and though they don't have to worry about getting along with crew, they still live in a close-knit, interdependent community where both fun and conflict are prevalent.

Then there are the families we meet on sabbatical for a year or two, on a trip they assumed would be like an extended vacation. Sunsets, fruity cocktails, dolphins leaping, white sand beaches, snorkeling in crystal clear water: these are the things that fuel their dreams. Annoying habits, lack of privacy, bickering siblings, marital conflict, seasick crew members: these are the stark realities of a group of people living together in a small vessel on the ocean. The same people who got along just fine in a climate controlled house—going to work, school, and social activities in the evening—discover that being with each other all day, every day in a space that is often short of comfortable, presents new relationship challenges.

Even those families who have decided to make living aboard a lifestyle, and not just a trip, develop coping mechanisms so that everyone can find equilibrium. For example, my friend April has been living on *Lark* with her husband and two daughters for several years. She has discovered that getting off the boat periodically to explore inland or fly back to visit family improves how she feels about the lifestyle, so they budget for family or individual trips. Other families, like our friends on *Water Lily*, live aboard seasonally—they live and work in Alaska half the year, and travel on their boat in the tropics the other half. It seems to prevent burnout and maintain excitement for travel. On *Take Two*, we try to balance a season of island hopping with a season at a marina so that we have the alternating benefits of travel *and* connection to a community. There are pros and cons: sailing offers us freedom and change but can be relationally intense, and being tied to a dock limits our movement but gives us easy access to shoreside amenities and opportunities to make friends among the locals.

However large or small the boat and crew, or however temporary or permanent the situation, living in close quarters reveals a lot about personality, relationship, and communication style. It starts with the two people who set out together—the captain and first mate—how they navigate their roles in the relationship, and extends to the rest of the crew as they divide work, find fun, and settle disputes. On *Take Two*, there are no formal articles of agreement or pirate code—only the messy process of disparate individuals figuring out how to get along one day at a time in a tight space.

Jay is the captain of *Take Two*—we decided never to argue about this. We've seen boats where there are co-captains, and we don't think this would work for us. In my opinion, a chain of command on a boat is important, and I have no real desire to be in charge. We make decisions together, but Jay's got the final say.

This is good because he is a reasonable person; when he raises his voice, it's to be heard, not because he's angry. I'm the impulsive one; I think out loud, argue, get emotional, and yell when stressed. I find his calm calculation both comforting and irritating. He solves problems like a computer: he sends his thoughts through a flow chart, comes up with a solution, and gives me the output without any explanation. While I can follow these orders, he doesn't verbalize the process so that I can learn to troubleshoot on my own. On the other hand, when he's struggling to solve a tricky problem, I may suggest a zany answer that he hadn't considered. He may find my monologues annoying, but occasionally, a good idea or two comes out of all the noise.

Our personality differences complement each other: both my exuberant spontaneity and his steady thoughtfulness are necessary to make our life work. Marriage is nothing if not the chance to experience, understand, and appreciate someone's otherness—we may start out at extreme ends of the spectrum, but ideally we meet somewhere in the middle. Without compromise, a couple walking the tightrope of marriage while balancing work, family, home management, travel, and a social life can quickly be thrown off-balance. That's true for any couple—but on a boat, it's even more important that two people work together. There's nowhere to escape when things are in turmoil. We are forced to resolve problems, communicate in a way that adds clarity and doesn't injure the other person, and attend to relationship maintenance… or sink the ship.

If that were not enough, we are also parenting five other people with strong personalities and in various stages of maturity. When we bought the boat, Eli was seven years old and learning to read. He's imaginative, responsible, and reserved, though easily-frustrated. He likes a well-ordered environment and can't stand large groups of people. Aaron was six, talkative and friendly, with a passion for anything having to do with tools, machines, engines, or wheels. He's ambitious and opinionated. Sarah was four, a girl equal parts princess and tomboy. She read early and showed a talent for music, but also took to sailing like a fish to water. She is smart and funny, but shy around strangers. Sam was fifteen months old when *Take Two* arrived at the dock in Bradenton, and he was walking, babbling, and into everything. He had a fun loving personality, swam early, and adored bird-watching and making friends with the other boat owners' pet dogs. His enthusiasm and energy can be irritating in a small space. To that mix, we later added a baby, a tiny person with a big personality, who would not have to make the adjustment from landlubber to seafarer, but who would rock the boat nonetheless.

Though we have seen each of them grow and learn in the decade since we moved aboard, their essential character qualities haven't changed much. Our children have strengths and weaknesses, like everyone else on planet Earth, and while the boat gives them a chance to exercise their strengths, the weaknesses are sometimes exacerbated by the small space. With every person added to a mix comes different chemistry and the dynamics change any time someone enters or leaves the room. Jay and I, already two opposite personalities with different approaches to life, took these chaotic little people and tossed them into an environment with constantly shifting variables. At the beginning, it was like herding cats—maybe fun, but nearly impossible.

The first year we owned the boat, we baby stepped ourselves to a level of comfort with packing up the house Friday afternoon, heading to the marina for the weekend, working on the boat or going for a sail, packing up again Monday morning and driving home. That meant homeschooling three days each week and car-schooling or boat-schooling the other four days. I was responsible for keeping a first grader, kindergartner, and precocious preschooler occupied and organized while juggling a toddler and his tornado of messes, diapers, frustrated efforts to communicate, and sometimes-irritated siblings—all while learning to sail, doing laundry, cooking, and cleaning. Jay, meanwhile, had to work a full-time job, spend time with the family, and go over every inch of *Take Two*, learning the systems and equipment. It was sometimes a bumpy ride.

One weekend, we forgot the bag that had the family's clothes in it. We were unpacking the car into a dock cart to wheel down the ramp to the boat. A few fat drops of rain began to fall as realization dawned on me.

"Honey, where's the grey duffle bag?" I asked.

"I don't know. It's not in the back of the car."

"It's the one we packed for the weekend. The clothes for everyone."

"Well, it's not here. Maybe we left it at the house."

"What are we supposed to wear?" I feel the pitch of my voice rising. "How could you forget the bag of clothes? All we have is what we're wearing! Are we supposed to go naked?" Not that it was his sole responsibility to remember everything, of course, but blame is an easy out, and the first sign of an irrational freak-out.

"Are you thinking of driving an hour each way to go back and get it? I'm not. I don't see why it's such a big deal. You have to go to the grocery store anyway. Just go find each of us some underwear and a couple of outfits."

"Easy for you to say! They don't sell clothes at the grocery store!" I'm close to shrieking, and the rain is falling faster. The kids are antsy and ready to walk down the dock.

"Tell you what. Get clothes, or don't get clothes. You can stand here in the rain if you like, but I'm taking the kids to the boat."

After throwing this small and pointless tantrum, I got back in the car and went to a department store outlet to get clothes and to the grocery store to get food for the weekend. By the time I got back, I had calmed down and the kids were playing happily while Jay worked on a boat project. He was right. It was not a big deal, unless I made it one.

There were countless other irritations, all of which could have been molehills turned to mountains. When it rained, the hatches leaked, and Sam found amusement in running around tipping over the cups and bowls we put under them to contain the drips. Every system on the boat needed repairing or replacing, a fact we discovered as parts broke, one by one, sometimes faster than we could fix them. We learned how to use, and troubleshoot, the dinghy and motor. We made mistakes while learning to dock and anchor the boat. We had mishaps while sailing. In our first year, we handled a haulout, a hurricane threat, and countless small emergencies. And we didn't sleep very well, so we were always walking close to the edge of insanity. No wonder people had told us it was a crazy dream!

After a year's worth of weekends on *Take Two*—digging through all the nooks and crannies, cleaning, fixing, and learning how to use the boat, we thought we might be ready to move aboard. But we didn't want to get overwhelmed, so we carried out a month long trial in April of 2009. We even brought Sugar and Spice with us—the same cats who had moved with us from Atlanta and who had begun to demonstrate their displeasure about our weekends away by damaging furniture and carpets while we were out. The experiment went so well that we decided it was time to put the house up for sale and make the transition to full-time living aboard.

Moving onto the boat with our family meant tackling several hard questions, one by one. *How would we pare down everyone's belongings and where would we store all the remaining stuff? How would we feed everyone out of a small boat galley? How would we educate our kids once we started traveling? How would we continue to earn enough money to pay for food, books, clothes, repairs, and everything else? And how would we keep everyone on board—both literally and figuratively?*

It turns out that getting rid of the stuff was the easiest part. We donated books to other homeschoolers and to the local library. We donated clothing and household items to local charities. We had a big garage sale and allowed the kids to keep any money they made from selling their stuff. We were surprised to find that they would rather have money than almost any toy in their collection. I gave each kid a box and said they could keep only what fit inside; they complied without complaining. For clothing, we kept only two weeks' worth each. For me, the hardest part was sentimental items—we still have a small airconditioned storage unit with wedding albums and baby photos, a few boxes of leatherbound books, and a collection of china tea cups I couldn't part with. After gathering necessities for the boat, the amount of extraneous junk we were storing in our house amazed us. Even now, in the pared down version of our life, we still have more than we need, and more than many of the people we meet (boaters and islanders alike).

The first week of August, 2009—five years exactly from when we relocated to Florida—we moved onto *Take Two*. At the time, Jay was engaged in an intense project at work and couldn't break away to help, so to be accurate, he stayed on the boat and concentrated on his job while the kids and I moved aboard.

To be clear, this was not packing up for a weekend on the boat, or a Thanksgiving trip, or a month aboard during cool spring weather; this was a backseat full of kids in the car every day for a week, driving back and forth between the house and boat, saying goodbye to our neighbors and our land life, and sorting through things from two households at the same time, in August in Florida. *In August, in Florida!* And then the air conditioner in my minivan broke. Initially, this didn't seem like a big deal to me—sure, it would be hot. *So, we'll roll down the windows. We're tough sailors now, no problem.* Unless it's raining. Every afternoon. (August in Florida, remember?) Then, a lack of air conditioning makes your commute a ride through the steamy Amazon, minus the snakes. I was driving with a load of kids and stuff, making stops to drop things off in storage, or to donation centers, swiping at the fogged windshield in front of me with a napkin from a drive-through restaurant while rain pounded down and a toddler cried in the back seat. Remember those Calgon Bath Powder commercials—*Calgon, take me away*!? We fixed the air conditioner and moved the rest of our belongings onto the boat.

A few months later, empty of children and all their detritus, the house sold and the last load of stuff went onto a Salvation Army truck (good riddance!) and we never looked back. This illustrates an important aspect of circumstances that cause conflict: they're temporary. Getting worked up and grumpy, taking anger or frustration out on the people around us, lashing out when we're hot, hungry, or tired, all these things can turn a temporary situation into a permanent relationship disaster. Remembering that whatever we're going through is short-term, but the relationships with the people with whom we're going through it might be lifelong, should keep us on a more even keel. I'm not sure I learned that lesson during our moving week, but I can certainly see it looking back. Thankfully, the children were so small that they have almost no recollection of this time in our lives. Grace takes many forms—forgetting being one of my favorites.

Once we had moved all our physical belongings, we had to

address the remaining questions one by one as they arose. Feeding everyone, for example, turned out to be simple, but not easy. I love to cook, but it is even more time consuming on the boat than in the house. Shopping for ingredients, storing food, preparing meals, setting the table, eating, and washing the dishes by hand: all of this takes an inordinate amount of our time, space, and energy. But, of course, eating well satisfies more than a physical need, so it's worth more effort. In the beginning, I planned and prepared every meal and snack we consumed, soup to nuts. As the kids grew and learned to help in the galley, they gained some autonomy—Sarah is our resident cookie baker, for example. We now have a rotation for galley helpers: meal preparation, setting and clearing the table, washing, rinsing, and putting away the dishes are a team effort.

If only running a homeschool were like cooking dinner! Take a few ingredients, throw them together, add some energy—voilà! Smart, capable humans who can read and discuss good books, do math in their heads, write an essay, understand the scope of human history and the innerworkings of the planet from atoms to zoology, speak another language (or two!), and possess some kind of skill in art or music or technology—this is a noble goal! But how to go about it? Homeschooling, or in our case, boatschooling, is part and parcel with living aboard and traveling. Before the kids came along, I was an elementary school teacher, but this, ironically, didn't really help much with teaching my own children. I have familiarity with different learning styles and with curriculum options, but motivating and inspiring my own little people to turn out of comfy cabins and get to work is not the same thing as showing up in a classroom, teaching a few hours, and sending my students home on a bus. In some ways, it's so much better—the curriculum dovetails with real life in a way that makes learning meaningful. In other ways, it's so much harder—

the lines between home and school, mom and teacher are blurred —and the one-room schoolhouse does quadruple duty as it becomes home, vehicle, and office too.

While I'm keeping everyone fed, clothed, clean, and organized, Jay is working from home. To call his space an office is a gross overstatement—it is a cabin in the port hull with a desk just big enough for his laptop and a floorspace too small for a proper office chair. Some families go sailing as part of a sabbatical; they usually save up a large chunk of change with which to travel and then return to work and normal life when the money is gone. Those families like ours, who have decided to live full time on a boat, must either find a job that is not brick and mortar, or split their time between working and traveling. So Jay, computer consultant extraordinaire, continues to solve technology problems for companies large and small from whatever port in which we find ourselves. In order to tune out the chaos in the environment, he uses noise-cancelling headphones and music, but, of course, that doesn't help much during conference calls. I can only imagine what his clients must hear in the background. We also have to plan travel around his work commitments, not just the weather. His job makes our life possible, even as it adds a layer of complexity.

With all these moving parts—seven people living, learning, traveling, working, and playing—sharing the same small space, there are bound to be conflicts. It's like a science experiment with too many reactive chemicals, and the explosions seem to happen at times of abnormally high stress, like when docking or anchoring, or while fixing a toilet. Or when someone is hungry or tired. Or when it's hot or the anchorage is rolly. Or when it's raining. Or while doing the dishes. Or on Monday mornings. Or Tuesdays, or Wednesdays…or…whenever.

So how do we make it work? With perseverance and grace.

Day in, day out, never giving up or giving in, we work at solving problems peacefully, and we offer each other mercy, grace, and forgiveness. In Shane Claiborne and Jonathan Wilson-Hartgrove's book, *Becoming the Answer to Our Prayers: Prayer for Ordinary Radicals,* the authors argue that "one of the most radical things we do is love the people we live with, day after day, mistake after mistake." It can be a loud and messy process, and we're still not very good at it. But it's also not as bad as it sounds. We often get along famously, we're productive and happy, and we laugh, play games, and make memories in beautiful places.

We are what you might call a *functional* family. When people express amazement that we live aboard a boat with our children, they are imagining, perhaps, what it would be like if *The Simpsons* moved onto a yacht. "How have you not killed each other?" they ask. Many of us come from *dysfunctional* families, so this is not hard to imagine. Living in a confined space and experiencing daily life together (instead of scattering in different directions) has given us lots of opportunities to cultivate the opposite. It is not that one family has problems and another doesn't, but rather how the family responds when problems arise. In dysfunctional families the *modus operandi* might include blaming, ignoring, numbing with alcohol or medication, walking out, name calling, screaming, guilt-tripping, shutting down communication, and generally repeating patterns from previous generations of dysfunctional families. With humans involved, there will always be a degree of this kind of dysfunction, especially the kind we inherit (either genetically or behaviorally) from our parents. But functional families address the problems, talk about them, listen to each other, come up with possible solutions, forgive each other and offer each other grace—chances to try again, to start over, and to repair relationship damage.

In essence, offering someone grace is giving them a second

chance—a "take two" (how fitting). On our boat, it works like this: someone comes upstairs grumpy and asks, "What's for breakfast?" I would respond, "Let's do a take two. Go back downstairs and come up again and try starting with 'good morning.'" When two kids who are doing the weekly sweep-and-mop bicker relentlessly during chore time, I might say, "We're going to do a take two. Tomorrow, you're going to do the chore again, this time peacefully. And if that doesn't work, you can do another take the next day, and every day after that until you are properly motivated to do the chore not just well, but *cheerfully*." This kind of offering a second chance may sound like a punishment, but it's really an opportunity for redemption.

When young children have an argument, they can be persuaded to apologize, make peace, and try again. They have short attention spans and seem to forget grievances rather quickly. But as they get older and have more life experience (and sometimes a pattern of victimizing or feeling victimized), saying, "you should apologize to your sister" or "you should forgive your brother" doesn't always fix the problem. Having a complete understanding of their responsibility when they have done wrong, and what to do when someone has hurt them is an important part of becoming a decent human being. When someone apologizes, he is saying, "I take responsibility for my actions which hurt you, and I'll try not to do it again." When the other person forgives, he's saying, "What you did hurt me, but I choose not to seek revenge, nor to carry this hurt around." It releases the person who hurt *and* the person who did the hurting. This process requires humility—a rare and desirable quality often lacking in humans.

We try not to leave unfinished business when there's been a disagreement. Jay and I abide by the rule, *do not let the sun go down on your anger,* in our marital disagreements, so we have tried to model conflict resolution and forgiveness for our kids. It's not that we don't argue—we do—but we don't do it with the intent to injure. We try not to hit below the belt, and we don't wait days to apologize or forgive. If someone has apologized for a wrong and the other has forgiven, then that wrong is not allowed to be dragged into a later argument. Regardless of differing opinions about things like child-rearing, gender roles, communication styles, and how to accomplish a task, we cannot afford to forget that we are on the same team, working toward the same goals.

If we didn't have a solid marriage, there's no way we could live on a boat with our children, so we place a high priority on marriage maintenance. When the children were small, they had a strict 8:30 bedtime, not just so they could get enough rest, but so we could get a break from parenting! We have always had date nights, but with a large family and the added difficulty of boat life, it's a lot harder to find a babysitter or to get away for a night. Sometimes date night is a sunset drink on the foredeck—no kids allowed. Sometimes it consists of dinner in the cockpit for two while the kids eat at the inside table while watching a movie. Having older children means being able to get away for short periods of time to go out for dinner or take a sunset dinghy ride. Keeping in touch with each other—who we are as a couple—is important for another reason: we're essentially dreamers, and dreamers need time to dream. Someday the kids will grow up and leave the boat, and the two of us will need to pursue dreams of our own again.

And while getting away as a couple isn't easy, allowing each adult some "time off" has always been one of our coping strategies. I read a book called *20-Minute Vacations* by Judith Sachs which inspired me to set aside a little time for myself during the week to do something relaxing. I have a hammock that hangs from our arch over the water between the hulls at the stern of our boat. It's hard to get into and equally hard to climb out of; once I'm in the hammock, the kids must figure things out for themselves. A paddle in the kayak or walk on the beach serves the same purpose. We can each ask for time off, to go out with friends or pursue a hobby or class or project while the other parent takes care of kids. That's one way I found time to write over the years.

Everyone in the family needs a little time to follow individual pursuits, and one benefit of a large catamaran is that there are cozy nooks where people can go to be alone. Whether Sarah is working on an art project in her cabin, or Aaron is practicing guitar in his, we've carved out little spaces for ourselves so we can share the boat and reduce conflict. As the kids have gotten physically larger and their hobbies louder (electric guitars, a bass, and drums, oh my!) this has gotten harder and harder. I wish I could say that we ironed out all the wrinkles in the early years of living aboard, but as people are always growing and changing, new wrinkles appear. It requires that we be flexible and unselfish —a tall order for family members, but skills that will serve us well in any environment.

A family is like a piece of braided line—when circumstances create chafe, the relationships are strands that can be broken, and the whole family will be compromised as a result. Communication is the most important tool we have to prevent chafe. We want everyone to feel listened to and cared for. When we notice a problem, we call a "family meeting." Anyone can bring something up for discussion, though usually it is an adult who has noticed a negative pattern or bad behavior. Sometimes the discussion happens at dinner time, but it can happen any time something goes wrong. Sometimes the discussion stretches into several sessions over long periods of time, especially if it's a recurring problem (like using kind words). Occasionally, we have private talk with one child after the rest of the family has gone to bed. These conferences can become tearful as people reveal private struggles—and end with hugs and prayers (especially when we aren't sure how to solve a problem).

Our hope is that by talking we can bring issues to the surface and keep emotional wounds from festering. The very purpose of a family is to give a person a place to share struggles and celebrate successes. Our family members are literally *all in the same boat* and are forced by proximity into this kind of closeness. What we've learned is that without a foundation in grace and forgiveness, good communication, and a willingness to let things go, we could never live in such close quarters.

# 7

## CHOCK-A-BLOCK
### COLLECTING VERBS

*December 2010. We wake up on Christmas morning to a calm and sunny day in the horseshoe-shaped bay on the north side of Highbourne Cay in the Exuma islands. It is our first trip through the Bahamas, and we came south from the Abacos only a few days ago on my thirty-sixth birthday. We stopped in Spanish Wells so I could go ashore to buy holiday supplies, and I found something special at Pinder's Market with which I can make a Christmas dinner. What we do not have this year are presents. Little by little, we have been whittling away at the "Hallmark Holidays," rejecting the commercialism of our culture and trying to find an authentic way to celebrate what is really holy: life and love and faith. It is hard to convince children that "money cannot buy happiness" when we have showered them with presents on every birthday and holiday of their short lives. Our own inconsistency has sent mixed messages. But moving onto the boat has forced us to minimize and reject the "more is better" mentality. Quite simply, we don't have space for more stuff. But will our kids feel disappointed? This is our first Christmas away from grandparents, and there are no presents under the tree. Furthermore, there is no tree.*

*Our inquisitive oldest child discovered early that "Santa" was just a game that grown-ups play, so we have never put on an elaborate deception. The children know who stuffs the stockings with goodies, and they assume that since we are far from Walmart and Target, there will be no treats this year. But they are mistaken. The smell of my childhood holidays, my mother's recipe for homemade cinnamon rolls, begins to waft out of the galley. The teakettle whistles and I pour hot water over the grounds in our coffee press. When the kids come upstairs, they are surprised to see bulges in the stockings (one of the only traditions we have kept and thought to bring with us). I have been secretly hoarding their favorite sweets since we provisioned in Florida in early November, not knowing what I would find in the Bahamas. Lucky I did, too, because the opportunities to shop have been few and far between. Tiny island stores, some no bigger than walk-in closets, are only well stocked on the day the mailboat arrives from Nassau. And I harbor one more secret: a brand new jigsaw puzzle I hid in my closet to bring out for a special day.*

*Everyone "ooohs" and "aaahs" over their unexpected Christmas candy. We have brunch and scrape up every hardened drop of icing from the baking dish the cinnamon rolls used to occupy. We read the story of the birth of Jesus from the Bible and listen to Handel's Messiah. We break out the new puzzle. And when the sugar high begins to peak and the kids are bouncing off the walls, we announce that we're going snorkeling. Imagine that—snorkeling on Christmas day!*

*At the northeast end of the bay are some scattered coral heads in six to ten feet of gin-clear water. It's called "The Octopus's Garden" on our chart, and with the gear in the dinghy, we motor over to the reef and look for a sandy spot to anchor. The water is so transparent and the colors so bright we almost don't need masks to explore below the surface. We spend the afternoon meandering over and through the coral gardens, searching for the elusive octopus.*

*Later that afternoon, we get enough of a cell signal to make a phone call to family in Florida. The kids talk to their cousins and I dread the old "What did you get for Christmas?" conversation. Instead, I am pleasantly surprised to overhear how happy our kids are with their day—eating sweets, doing a puzzle, and snorkeling with their family. I feel triumphant: our kids don't need stuff to make them happy! If they realize this now, imagine the contentment they can experience their whole lives, and how little time and energy they will waste pursuing worthless things. For me, it is the best kind of gift, a day full of memories that will never end up in a landfill.*

---

Despite getting rid of ninety percent of our belongings when we left our landlubber life, our boat is still chock-a-block. When we moved onto the boat, Jay threatened to weigh the boxes I brought on board; he was alarmed at how the waterline crept up as I stowed school books, clothes, kids' toys, dishes…even cast-iron skillets! *Chock-a-block* is an old nautical term that means tightly packed—imagine a cargo hold crammed so full that nothing can move. Go on any liveaboard boat and ask to see a storage locker or lazarette and you will see what I mean. Stowage under settees, stowage in the bilge, stowage under berths: every space is jammed *chock full*. When you want to get something out, you must remove most of the items, find and retrieve the one you want, then carefully pack everything back in the way it was. Before a passage, all the loose items must be tucked away in those lockers because they can become projectiles in rough seas.

But there's another sense in which our boat is chock-a-block. We have filled it with keepsakes, experiences, and memories. One of my favorite children's picture books, *We Were Tired of Living in a House* (by Liesel Moak Skorpen and illustrated by Doris Burn), depicts the children moving out and slowly swapping the belongings they brought from home for mementos from their treehouse, raft, cave, and beach. Our boat is similarly full of odds and ends, seashells and interesting rocks we've picked up on distant shores, souvenirs, tiny paintings from local island artists, and currency from other countries. Even the pantry contains weird foodstuffs from foreign markets. These are the *objects* we've collected, but like passport stamps, they are mere reminders of the *memories* we've gathered.

When people buy a boat and sail off into the proverbial sunset (or sunrise, as is more often the case), they are likely imagining dolphins leaping in their bows as they sail across azure seas, ghosting along under full sail in a light breeze at night under starlit skies, catching fish for dinner while underway, drinking sundown cocktails on deck, and anchoring in exotic places with island views and coral reefs in the liquid backyard. They look forward to making friends everywhere they go and connecting with fellow travelers who are also seeking a life of adventure. They may even imagine storms at sea, evading pirates in high-speed boats, and having to repair an engine or stitch a torn sail far from land and without help. In fact, cruising on a sailboat includes all of those elements. It can also be incredibly mundane, uncomfortable, and sometimes downright miserable. But even the unpleasant things make for good storytelling later, and so people like us have been willing to trade a comfy life ashore and a house full of objects for a simpler life afloat and a boat full of exciting memories. My friend Davina on *Island Girl* had a perfect saying for this: "collect verbs instead of nouns." And that is exactly what we have done.

When we moved onto the boat full time, we had kept only what would fit (or could be crammed) in the available storage spaces. We realized very quickly that traditional commercially-driven holidays weren't feasible—we didn't need any more stuff. If anything, we needed to continue to pare down what we already had. Birthdays and Christmas—the main gift-giving holidays—would have to change drastically to keep the waterline from creeping up any farther.

Our first chance to put our new ideas into practice was Eli's eighth birthday, which we celebrated at a bowling alley during our transition to living aboard. He and his best friend, Jonah, were born one day apart, and his mom, Tarin, and I (neighbors and inseparable friends) decided to host a joint party. There were balloons, a giant cake, friends from Jonah's school and our group of homeschool friends, grandparents, bowling fun (of course), and piles of presents. Presents for Jonah, anyway. We had told our friends not to bring anything material—that their *presence* was better than *presents*. The only gift he received was a preapproved looky bucket from Skipper and Grandma Mary, a five-gallon pail with a plexiglass bottom for seeing what's under the water's surface. The party itself *was* the gift, a memory that would last longer than the things that came wrapped in pretty paper. You might call this *birthday do* instead of *birthday get*. If we spend money, we want to have something to show for it that lasts, that prioritizes relational connections over object collections.

This first boat birthday was a harbinger of good things to come. Later that year, we would celebrate Aaron's seventh birthday by inviting his six cousins, along with my brother and his wife, to go sailing. We anchored off the beach at Eggmont Key at the entrance to Tampa Bay, grilled hotdogs on the back of the boat, let the kids swing like monkeys from our arch and jump off the boat into the water, fished, swam to the island, explored the coastal forest and the historic fortifications from the time of the Spanish-American War, and enjoyed a pirate-themed birthday cake as we motored back to the dock. Sarah's sixth birthday the following spring involved horseback riding and a cake decorated with small toy horses prancing on grass made of coconut flakes dyed green. And Sam's fourth birthday would be the first of what we call *Bahama birthdays*, marked by inviting other cruising friends over to play on the boat and swim in the turquoise water. When we ask what our children would like to *do* for their birthdays, it gives us a window into their passions and interests and allows us to share those experiences with them. It communicates respect and genuine interest, and it represents a sacrifice of time, effort, and resources that sometimes exceeds the cost of a material gift.

Over the years since these early days, we have fulfilled many birthday wishes: a trip to a climbing gym where Eli learned to rappel, a visit to an air museum, where he got a ride in a biplane, go-karting and ATV-ing with our motorhead son, Aaron, scuba diving and Hobie sailing with Sarah, a trip to Legoland with Sam and a camping birthday where he blew out flaming marshmallows instead of candles on a cake. Traveling would make the birthday outings into cultural experiences and opportunities to explore the natural world. Imagine, for a moment, that instead of collecting toys, your child could gather memories like climbing a volcano, hiking in the jungle, surfing in the Pacific, snorkeling on a reef, or jumping in a waterfall. These are the *gifts* we had always hoped to give our children, and the boat made it possible.

Our parents were a little slower to catch on to this new way of doing things. Naturally, they wanted to continue spoiling our kids—they view it as their reward for raising children, saying *yes* as special right reserved for those who spent a lifetime saying *no*. We tried to tell them *no holiday gift-giving* when we moved onto the boat, but they were not happy about it. One year we convinced them to donate the Christmas money to a charity that provides small villages in impoverished parts of the world with livestock. Another year we asked them to help us buy kayaks instead of individual presents. Once, Jay's mom and stepdad insisted on buying gifts, but decided that the toys would stay at *their* house to be played with on visits. Another year, my dad planned a Christmas outing and took our family and my brother's family to a water park for the day—a memory we still cherish.

Finally, after they had accepted that their gifts couldn't take up space and we had accepted that they still wanted to spoil their grandchildren, we hit on an idea that has stuck: *the Fun Fund*. Instead of giving us stuff, the family sets aside money for use in making a memory—turning the gift from a noun into a verb. One of our favorite uses of these holiday cash gifts is travel. In the years since we first started it, the Fun Fund has taken us high and low in search of memories: skiing in Maine, hiking in the cloud forests of Panama, surfing in Costa Rica, exploring the old walled city of Cartagena, and horseback riding among the volcanos of Guatemala. It is a win-win: the grandparents are happy to see the pictures of the children on their adventures, our kids have enjoyed unique activities, and we don't have to store the detritus from a decade of holiday giving. I'm not saying that Nerf guns and Barbies aren't fun to have around, or that our kids don't have plenty of toys, games, art supplies, and musical instruments to keep their hands and minds busy, but whereas Christmas or birthday gifts lose their luster after a few weeks of use, people never outgrow travel and memories become a permanent part of who they are.

In essence, this is why we bought the boat in the first place: to travel with our children. By our third year of boat ownership, we had learned to deal with the lower lows of breakage, discomfort, and isolation, and we had begun to experience the higher highs of a thrilling downwind sail with the spinnaker flying, anchorages near little islands where we could let the kids run wild, and quiet evenings out on deck under a moonlit sky with a glass of wine in hand. We had seen wild weather, shore birds, sea creatures, and sunsets that looked like molten metal poured onto the sea. And we had made a promise to our kids: "One day," we said, "You'll go to sleep in one place, and wake up in another."

We didn't have to go far or wait long to fulfill this promise. Without fanfare, without mishap, we motored away from the marina one morning in June of 2010, sailed out the mouth of the Manatee River, and entered the Gulf of Mexico. We weren't sure if we were coming back—whether this was just a trip or the beginning of a traveling lifestyle. We just knew it was time to go. We had sold our house, acclimated to boat life, potty trained our toddler and taught him to swim, and learned as much as we could about sailing, but we had never taken our family on an overnight passage or spent time offshore. We had been caught in afternoon thunderstorms but had never faced any real danger. We didn't know how our children would handle long sailing days with no chance of going ashore. I had never had to cook while the boat was rocking in ocean waves. Jay had never had to fix something that broke while underway. We set off—aware of all the things that could go wrong, but hopeful that we could figure them out as we went.

We pointed the boat toward the waters of the Florida Keys, where Jay's family had taken sailing vacations during his childhood, and where we had taken our trip with his parents as newlyweds. For our first overnight sail, we instituted a three-hour watch rotation. As the last view of land disappeared over the horizon, we took care of evening routines: dishes, brushing teeth, showers, and putting kids to bed. I then went on night watch at eight o'clock with a cup of coffee and a good book. Three hours later, I woke Jay with his cup of coffee, and he took a watch. After what seemed like only moments of rest, he woke me (now two in the morning), and I groggily sat in the chair, looking up into an unpolluted, dark, star-strewn sky. At five, more asleep than awake, I swapped places with Jay again. We arrived in the Dry Tortugas as the morning sea planes touched down, like ducks landing on a lake, to deliver the first visitors to Fort Jefferson on Garden Key. As we set the anchor in the little bay near the fort, the kids poked their heads out of hatches like groundhogs waking from hibernation. You could see the words written across their faces, "Where are we?"

Though Jay and I felt a bit punch drunk—equal parts giddy with success and groggy with fatigue, our kids, who had slept through nearly the entire trip, were not tired at all. They were up and ready to go swimming, snorkeling, exploring, and fishing. We made breakfast and an extra pot of coffee—the elixir of life—while the children put on their swimsuits. After a quick bite, we donned masks and snorkels and jumped in. We had never seen water so blue or so clear. Right away, the children began scouring the area for wildlife. They would snorkel around the boat, popping their heads up intermittently and shouting excitedly.

"Look, Mom! There's a starfish on the bottom!"

"Dad, there's a barracuda under our boat!"

"Mom, do you see those black and white striped fish? They look like zebras!"

"Whoa! Did you see that? It was a nurse shark! It swam right under our boat!"

Every day held surprises. We toured the fort, completing the scavenger hunt in the Junior Ranger booklet, where we learned about the fort's most famous occupant, prisoner Samuel Mudd, the doctor who conspired with John Wilkes Booth and treated his broken leg after he assassinated Abraham Lincoln. We snorkeled on our first reef on the back side of Loggerhead Key, amazed at the shoals of colorful fish that floated like clouds of silver, yellow, and blue above the hills and valleys of coral. Even Sam, who had just turned three, donned miniature fins and a small mask and snorkel. He happily floated face down in the shallow water, holding my hand and watching as tiny, candy colored reef fish darted in and out of holes in the rocks. Jay took the older kids fishing, and returning in the dinghy, he rescued a couple whose inflatable was sinking, the tubes having delaminated from the fiberglass bottom. We slept out under the stars one night and were awakened at dawn by the sounds of a colony of sooty terns and their hatchlings beginning to stir. When a passing thunderstorm threatened us with rain, we closed all the hatches and went out on deck for a free shower. At the end of a week, we sailed for Key West.

The Florida Keys have their own special vibe. A chain of coral limestone islands surrounded by blue-green water, they stretch about a hundred miles west from the mainland, connected by a series of bridges built by Henry Flagler at the turn of the century to carry Overseas Railroad passengers. This first wave of Gilded Age tourists could travel by train from New York to the port of Key West, where they could board ships bound for Havana and the western Caribbean. This railroad, and the highway that later paved over the rails to carry automobiles, shaped the geography and history of these subtropical islands. Jay and I remember the Keys from childhood trips with our families (his arriving by boat, mine by car), before the fancy resorts were built and cruise ships stopped in Key West, when the islands were quiet, charming, and historic. The bustling, touristy town we found when we pulled into the Conch Harbor Marina in Key West twenty years later looked like an antique that had been given a coat of gawdy paint. It was recognizable, but different.

After the weeks of sailing, living on the anchor, and

handwashing clothes, staying at a marina was a luxury we had not properly appreciated before. I sat by the pool with a piña colada, watching the kids swim, while a machine in the marina laundry room magically washed my clothes *for me*. We closed the hatches and ran the air conditioner so we could sleep at night without sweating or swatting at mosquitos. We all took long, hot showers at the guest shower house. When Al and Mary flew down to join us for a few days, we walked all over Key West together, stopping at the attractions children might enjoy—the Wreckers Museum, the aquarium, the Southernmost Point, the Butterfly Conservatory, and the Hemingway House with its gardens crawling with six-toed cats, descendants of Earnest's own odd feline, Snow White. We had lunch at an outdoor café with chickens pecking in the yard, and in the evening, we went to Mallory Square, a long wharf lined with gift shops and restaurants, where one can watch not only the dramatic sunset but also street performers—fire eaters, jugglers, magicians, acrobats, and musicians. Though we enjoyed visiting Key West, after a few days, we were ready to move on.

Without a schedule, without the need to be plugged in because we could make our own power (with solar panels and a generator) and our own water (with a desalination unit), we were free to slowly meander our way along the chain of islands, stopping to snorkel on the barrier reef and anchoring for the night in cozy bays. When we arrived in Marathon and picked up a mooring in Boot Key Harbor, we had no idea of the significance this place would hold in our lives, only that we needed somewhere to settle for a few months so that Jay could work and we could prepare the boat and crew for a trip to the Bahamas.

This mangrove-lined bay provides reasonable protection from all wind directions and looks like a floating village, with hundreds of boats on evenly spaced mooring balls. It was here that we were first introduced to the cruising lifestyle and to the very special community of sailors and liveaboards who call the ocean their home. Living on a mooring in that safe harbor was like cruising with training wheels: we learned how to provision by dinghy, how to communicate on the VHF radio, how to pick up and drop a mooring, how to live off-grid, and how to make ourselves comfortable despite the heat, humidity, bugs, and other irritations. We learned how to manage our daily routines without the added challenges of weather routing, sailing, entering foreign ports, and learning a new language or culture—all of which would come later. And we made friends.

Other families cruising with children stopped by our boat in their dinghies or approached me in the laundry room and introduced themselves. They answered my questions, gave good advice, and commiserated. We also connected with other homeschool families living on land. Every Tuesday, staff from the community park would hold a P.E. class for the homeschool children so they could play sports, run out some energy, and meet friends. The homeschool moms would sit in the shade and chat, making plans for field trips, beach days, and moms' nights out. We were quickly embraced by this diverse group, a loose collection of people who came and went, traveled extensively, and lived alternative lifestyles. Some were permanent residents, and others seasonal visitors. This group of friends formed a kind of home base from which we would launch into the great blue yonder, and to which we would later return and be welcomed back as if no time had passed. They became part of the fabric of our lives, woven of people and places and happy memories.

And just when it seemed that life could get not get any fuller, we learned we were expecting another small crew member. Talk about chock-a-block! *Where would we put a baby and all the stuff that one requires? Would I be able to travel while pregnant? And how would a newborn change our lives?*

Sarah had been asking for a sister, telling me repeatedly that she was "tired of all these boys." A baby sister was the main focus of her daily bedtime prayers. Of course, there were no guarantees it would be a girl, and we hadn't exactly planned to add another person to the family, but we embraced painter Bob Ross's mantra: "There are no mistakes, only happy little accidents." We had managed with four little people, and somehow, we would survive the addition of one more. The due date was just before Sarah's seventh birthday in May, so we decided to use the winter months to travel while we still could. I made an appointment with a nurse-midwife to make sure everything was progressing normally, and as soon as we were through the first trimester, we provisioned the boat, filled the tanks with diesel, and studied the weather, looking for a window of opportunity to cross from Florida to the northern Bahamas.

We crossed the Gulf Stream from Florida to the Abaco Islands in November 2010, choosing calm weather for motoring instead of wind for sailing across this notoriously bumpy stretch of water for the first time. We listened to the drone of the engines all the next day as we passed over the Bahama Banks—an endless expanse of shallow, impossibly blue water. It was like looking at the world through a bottle of Bombay Sapphire Gin—we could see individual pebbles on the bottom and count starfish and sea biscuits as *Take Two*'s shadow rippled over the bottom. A welcoming committee of dolphins came to greet us, rolling on their sides to look up at us with their smiling eyes and playing in our bow wake. We anchored near Great Sale Cay at the end of the day and dinghied ashore to stretch our legs. We pulled our inflatable up on the sand, unbuckled the kids' life jackets, and they hopped off barefoot onto the cream colored sand of a

Bahama beach. They brought us treasures we had never seen before—the empty exoskeleton of a sea biscuit, a large pink conch shell empty of its inhabitant, a striped snail, algae that looked like lettuce leaves. We putted back to *Take Two* where she was anchored in water so clear that she seemed to be levitating above the sandy bottom.

The next day, Jay and I woke in the dark and pulled anchor before the children rose. Coffee in hand, we motored away from the anchorage and prepared for a day sail in brisk wind across the Sea of Abaco to Green Turtle Cay, where we could check in with customs and immigration and get our first passport stamps. Just as we had promised, the kids fell asleep in one place and would wake in another.

Since those first islands, there have been many landfalls in mysterious new places, many passages that ended, despite our best efforts to get there by daylight, in nervous approaches to unfamiliar channels in the dark, and shadowy anchorages we had to navigate by depth sounder and chartplotter. We have often dropped anchor in view of an unknown island silhouetted against starlit sky and then dropped ourselves, dead tired, into our bunks. Like the dispelling of an enchantment, the light of the rising sun reveals the hidden world to our curious eyes. Things that looked lumpy in the dark turn out to be sparkling mangrove islands, volcanic boulders, or local fishing boats. Water that looked inky black turns emerald, turquoise, or aquamarine in the morning light. Twinkly lights on shore turn out to be sherbet colored houses sprinkled along the beach, thatched huts emerging from a jungle, or fancy palm-studded resorts. We never know what the daylight will reveal—but it will never look like we imagined when we were reading cruising guides or flipping through tourism magazines. The real place will always simultaneously delight and disappoint, and then become part of our permanent collection of places.

People say you can't buy happiness, but I think that's only partly true. It really depends on how the money is spent. If it's used to bring people closer together, buy a little freedom from the daily grind, pay for memorable experiences, and help pursue dreams, then we would say, "spend it." That's what money is *for*. That's what life is *for*. What we ought to say is that *buying stuff* does not ensure happiness. We have rarely missed any of the things that we got rid of when we bought our boat. We regularly get the *urge to purge* and continue to get rid of things we don't need anymore, replacing them with mementos from our travels or adventure gear with which to make more memories. That's what a small space offers: the opportunity to find out what you *do and do not need* to be happy. It was liberating to pare down our belongings and free ourselves up for going places, meeting people, and doing things instead. Instead of impoverishing us, we've learned that replacing our nouns with verbs fills our lives chock-a-block.

# 8

## ALL HANDS ON DECK
### TEAMWORK

*February 2011. There is barely enough space for me and my growing belly in the grocery-laden dingy as I putt-putt back across Elizabeth Harbor from George Town, Exuma toward Sand Dollar Beach, where* Take Two *is anchored in the lee of Stocking Island in the southern Bahamas. I have grown considerably over the last three months since we left Florida, and Jay's prophecy about my discomfort while climbing in and out of the dinghy with groceries has proven to be true. When we found out I was pregnant, I had nonchalantly said, "Babies are born everywhere. We'll just stop off on an island somewhere, add a crew member, then keep going." He looked at me sideways and said he thought I might feel differently in six months.*

*Approaching the stern, I call "grocery brigade!" and Jay rings the ship's bell. All four kids—from nine-year-old Eli to four-year-old Sam—come running when they hear the all-hands-on-deck signal and crowd into the cockpit to help unload the dinghy. There was a time not so long ago when I couldn't imagine how I would manage provisioning when we sailed away from the dock and left our car behind. But after several months of living life unplugged, this is all normal procedure.*

*Going to the store used to take me an hour-and-a-half maximum but now requires half a day, plus recovery time. I find a time when I can hand off childcare duties to Jay (so someone can keep a watchful eye on those four monkeys), load my canvas grocery bags into the dingy, lower it from the stern arch into the water, motor across the harbor and into the protected bay behind town, and tie it up to the dinghy dock at Exuma Market. After carefully climbing out of the dinghy, I waddle my pregnant self up the dock in tropical heat and then make several stops in town, looking for produce, fresh baked bread, and groceries. I finish at Exuma Market, load up a cart, and check out. I wheel the grocery cart down the dock, load the bags one by one into the dinghy, return the cart, and waddle back, carefully stepping into the inflatable. I start the motor, back away from the dock, and exit the narrow channel. Crossing the choppy harbor, I am careful to take wakes from other boats head on, so I don't soak my expensive food stuffs with saltwater.*

*I tie the dinghy to the stern cleat and begin handing grocery bags up to the waiting helpers, who line up to pass items from the dinghy to the galley. I hand a bag to Jay, who passes it into the cockpit. The cockpit helpers hand the bags into the cabin, and the last person in the line begins to unpack bags and stow provisions while the youngest crewmember stumbles around carrying lightweight items and generally getting in the way. The whole operation takes about fifteen minutes, and then I spend another hour working on the jigsaw puzzle of food storage on a boat—organizing and stowing items in various lockers and cupboards.*

*What once was a one man job done by air-conditioned car between an air-conditioned store and an air-conditioned home is now a hot, sweaty workout (and sometimes a wet mess, depending on weather), and requires literally all hands—though it is not without its rewards. I dig a bag of melting M&M's out of the last grocery bag and compensate the crew for their helpfulness. The captain gets paid in beer. And I finally sit down in the shady cockpit with a cold limeade and put my swollen feet up. That will be all I have energy for today, but that one task is enough.*

---

We discovered early in our liveaboard life that despite the relatively small size of the boat, communicating when people were spread out stem to stern was not easily accomplished. I don't mind the shouting, but Jay finds it irritating (especially when he's on the phone with a client) and even embarrassing (when living in close proximity to neighbors). So we began to use the ship's bell to get the attention of the crew. We are fond of this brass bell, engraved with the boat name, year of launch, and place where she was built in the Netherlands: "TAKE TWO OFF T'WAAR, BOUWJAAR 1991." We polish it only sporadically, so it looks ancient and takes on a patina of beautiful sea-green oxidation when properly neglected. It hangs near the center of the cabin, within easy reach of the galley. The bell rope is a nice, thick braid with a monkey's fist at the end. The sound when the bell is rung is pleasant and carries far enough that kids out on the dock or down below raking through the Lego bricks can hear it. We decided to keep the requirement simple to get all hands on deck: *if you hear the bell, come to the salon.*

Often, the bell gets rung for mealtimes, so it takes on Pavlovian significance. The kids hear the bell and wonder if there's a smoothie in the blender, if it's time to set up for dinner, or if muffins have just come out of the oven. One time I rang the bell and called "Cantaloupe!" The boys came running from their cabin and said, "Candy loaf? What's a candy loaf?!" I barely had the heart to tell them it was *just* a melon. Imagine the disappointment when I ring the bell for all hands and ask kids to do their chores!

Chores figure heavily in our weekly routines. Keeping a small space orderly is necessary for productivity, creativity, and, of course, *sanity*. A family is a place where people learn the give and take of life—if you take, you must also give back. Those who want the fun of living on a boat must also learn how to keep that boat orderly. We ring the bell when the trash needs to go out, or when we need a volunteer to run an errand, or when it's time for the sous chef to help with dinner, or when we need a grocery brigade to line up and unload the dinghy.

We also ring the bell for a family announcement or meeting, or when it's time to go somewhere. We ring the bell when it starts to rain and we need everyone's help to batten hatches or bring in laundry. And, on rare occasions, we ring the bell for an emergency. Thankfully, those times are few and far between. But if we ever had a fire or serious injury, knowing that the kids would respond and spring into action gives me a sense of reassurance. This is the idea: the bell will ring and, magically, a group of rugged individualists will fall into formation and respond like gears in a machine, each doing his part to complete the task at hand. At least, that's how it is *supposed* to work.

In reality, teamwork is hard. People have different gifts and challenges. They have competing interests, personality traits, and priorities. With good leadership, a disparate group of people can come together to complete a task or mission much more efficiently than they could working as individuals. And while one might not describe our family as a well-oiled machine, living on the boat forces us to work together for a common purpose because daily life requires everyone's participation.

It is one thing to ask children to do chores as a way to teach a work ethic, or to make them feel valued. It is another to ask for help from all the family members when a task would be impossible without it. We always had a chore chart posted next to the boat rules, with jobs that rotated monthly, giving children the chance to get good at a task without getting bored with it. Sometimes we assigned jobs that had to be done with a sibling, like washing and rinsing the dishes, which encouraged partnership and teamwork. Other jobs came with an intrinsic reward: helping with provisioning meant choosing a treat at the store, sharing in meal planning and preparation meant making a favorite dish or dessert, doing the laundry meant having clean clothes. Often, I would offer incentives, promising a frozen lemonade after everyone had finished the weekly floor swabbing, for example. But when we started traveling, and then I got pregnant, chores took on even greater significance.

For years, while the kids were young, I had double the work; training a child to do a chore is twice as hard as just doing the chore yourself, but I knew I was making an investment that would pay off sooner or later. It turned out to be sooner. When we were island hopping in the Bahamas, we went days or weeks between laundromats, so we had to wash clothes by hand. It was nearly impossible to do this with my own two hands—we had a Wonderwash rotating bucket, which one person would crank to agitate the soapy clothes while another person would rinse the previous load and send it through the Dynajet wringer as I turned the crank, and a third helper would hang the wet clothes on the lifelines to dry in the wind and sun. Even Sam, who was little more than a toddler, would help by gathering the clothespins when I took dry clothes down at the end of the day. No one in our family takes washing machines and dryers for granted!

Travel days require another kind of teamwork altogether. A large sailboat is a complicated machine. From hoisting the anchor to raising the mainsail, from navigating to steering toward waypoints, from lowering sails to arriving in port and tidying lines, extra hands are always needed. Even those who are not actively participating in sailing can still be helpful. On passages, children started taking turns in the captain's chair at age eight to watch for ships and shoals and learn how to pilot and navigate, while an adult offered support. Those who do not suffer seasickness often help with food preparation or fetching things for those who do suffer. We post a lookout in shallow waters with areas of scattered coral heads or in coastal waters where the surface is littered with crab or lobster floats. Knowing that we can count on our kids to cooperate when they are needed, or to sit tight and behave themselves when they are not, gives us the confidence to venture out of our comfort zone and into the unknown.

By February 2011, we had decided that we should head back to Florida to have our baby for three reasons. First, though babies *are* born in every country in the world, in the Bahamas they are born only in Nassau and Freeport, where the big hospitals are. Island women go to these larger cities in their last month of pregnancy to wait for the birth of their babies. We were neither interested in staying in those ports, nor in having a baby in a hospital. We had visited island midwives for prenatal checkups along the way, even getting an ultrasound in Marsh Harbour, but they no longer deliver babies in local clinics unless it is an emergency. Second, I wanted to be near our support system of family and friends, most of whom were in southwest Florida. We realized we were in over our heads and would need extra help after the birth. Finally, adding a baby to a boat would mean making some renovations to our interior to create a safe place for a baby, and to generally improve the living space for a large family. By now, we were committed to our lifestyle and felt good about making some big changes. With eight weeks left before my due date, we sailed north from George Town to work our way up the island chain and across the banks.

We had chosen a weather window that would allow us to get ahead of an approaching cold front, anchor for a night or two in a sheltered bay while it passed over, and then continue on our way. It was cool, sunny, and breezy—beautiful sailing weather. We waved goodbye to our good friends on a neighboring boat in the anchorage, *Begonia,* with Karla and Sebastian and their two kids Sofia and Benjie aboard, not knowing if or when we would see them again, as they were preparing for an Atlantic crossing. We exited Elizabeth Harbor and headed east into the Exuma sound, waiting until afternoon to enter a narrow channel between islands to sail over the Great Bahama Bank, which would mean calmer waters. Or so we thought. We passed other boats tucked into cozy anchorages where people were swimming and fishing. We wondered if we should stop early, but we were making good time and figured we could make it as far as Little Farmer's Cay even if the weather arrived earlier than expected. *How bad could it get*, we reasoned, *on shallow banks?* The answer, as you may guess, is *worse than you think.*

With only a few hours to go to the anchorage, the wind began to shift and dark clouds piled up, seeming to block our way. The sea went from a sparkling turquoise jewel to a foaming gray maelstrom in mere minutes. Together, Jay and I furled the jib that had begun to flap; he winched in the furling line on the port side as I eased the sheet on the starboard side. We turned on the engines and I steered the boat upwind as Jay went forward to reef the mainsail. The waves became shorter and steeper, and spray began to hit the main cabin hatches. I went below to get Jay's foul weather gear while he steered us through the storm. With our captain at the helm, I sat with the children, who had gathered in the salon, waiting to see what would come next. We felt like we were going through an automatic car wash.

Suddenly, the port engine fell silent. I opened the door to the cockpit and told Jay something was wrong. I put on my own waterproof jacket and took a turn at the helm while he went to investigate. With the large volume of water on deck, seawater had leaked into a fuel fill port, and the diesel on that side was contaminated. There would be no more running the port engine until we could filter the fuel. We pressed on with the starboard engine running at higher rpms. I took off my dripping jacket and sat down with Sam on my lap, hugging him and reassuring him and the other three children that we had the situation under control. I spoke too soon.

Just then, an ingress of seawater in the water compartment shorted out the propane gas alarm, which had to be silenced manually until someone could go forward, open the hatch, and disconnect the alarm. It went off every few minutes, like a snooze alarm, blaring a high-pitched beeping that heightened the tension. I then heard a sloshing noise coming from below, and carefully made my way down the starboard companionway stairs to inspect. The ocean was making a valiant effort to get into our boat and seawater was dripping onto the mattress in the master cabin from the hatch above our bed and leaking by the gallon through the improperly sealed hatch above our shower. I tightly closed and turned the handles to dog the hatches and prevent further leakage and then made sure the pump evacuated the accumulated water. I came back upstairs just as Eli said in a tremulous voice, "Mom…you might want to take care of this. Sugar just pooped on the seat next to me." We had literally scared the crap out of our cat. I grabbed some paper towels, threw the mess overboard, and came back just in time to see the drawer of cutlery come flying open. I removed the drawer and set it on the floor. A second drawer slammed open. Soon, four drawers lined the galley sole. I flopped onto a cushion and said a prayer, out loud, as much to calm myself as to reassure the children that God still had us in the palm of his hand. "We got ourselves into this mess, but you can still help us get out of it," I prayed.

Jay opened the door and asked me for a glass of water. Just then, a monstrous wave swept over the cabin top, and I watched as thirty gallons of seawater poured onto his head. I quickly shut the cockpit door and waited for the water to drain through the cockpit scuppers. When I opened the door again, the acrylic cup he held was overflowing with saltwater. He dumped it out and said, "*Now*, can I have a glass of water?" I rinsed and filled the cup at the galley sink and noticed that something was wrong on the foredeck. As I handed him the water, I told him that the wave had broken the catwalk and a trampoline had torn loose and was being dragged underneath the boat.

I took the wheel again, trying to keep the boat steady so he could crawl forward on hands and knees with a knife to cut the trampoline free. When he told me to head up (as in upwind), I turned the wheel the wrong way, accidentally heaving to. The sail backwinded, the boat gave a lurch, and then everything grew strangely calm. It was as if we had pushed a pause button. Without forward momentum, the boat stopped bashing, steadying itself in a gentler rhythm with the waves. We both took a deep breath.

"What just happened?" I asked.

"I think we just hove to," he responded. "I didn't even know a catamaran could do that."

"What do we do now?"

"I'm going to get that trampoline in, and then we're going to take the main down and turn around. Look for an anchorage on the chart where we can shelter until this blows over. We're not going to make it to Little Farmer's."

"Got it. Be careful."

While Jay cut the catwalk free and dragged the sodden mesh trampoline back to the cockpit, I looked for a safe place to tuck in for the night. The children, meanwhile, sat silent and wide eyed, oddly calm. When Jay caught his breath, I showed him the chart I had been studying. We pinpointed what looked like a good place to anchor for the night and marked it on the chartplotter. I turned upwind just long enough for him to lower the mainsail and zip it into the stack pack before heading south again to retrace our steps toward an anchorage behind a small island. I came into the salon, told the kids the plan, and then sat down to read Beatrix Potter aloud while Jay ran before the storm.

Before dark, we had the anchor down, the wind and waves nearly disappearing in the lee of a small island. Relief washed over us. Nothing had broken that could not be mended, and we were safe. After the crashing and banging of the storm, the quiet lapping of water on our hulls was like gentle music, soothing to our anxious minds. We began to clean up—the kids and I indoors, and Jay on deck—and cook dinner, pretending that this was just another stop on a pleasure cruise. We washed dishes and put children to bed, then crept into our bunk earlier than usual, exhausted from the adrenaline filled afternoon. I whispered a prayer of thanks as I fell asleep.

By the next morning, the wind had abated, the seas were calm, and a grey blanket of clouds was all that remained of the fierce front that passed overhead. Jay used the fuel transfer system he had installed the previous year to filter the saltwater out of the diesel in the port tank so we could get the engine running again. I washed and hung salty clothes and towels as best I could. The cat and children seemed to recover quickly, relaxing and playing while we prepared for the next leg of our journey. Jay and I debriefed, asking ourselves, *What should we have done differently? What did we do right?*

In retrospect of course, it is easy to see that we should have stopped earlier or changed course immediately when the storm arrived. That cold front certainly changed the way we read forecasts and handled heavy weather going forward. It revealed the strengths and weaknesses of our vessel, and added a few things to our project list. It also taught us something else: teamwork is a prerequisite to sailing in a storm with a boatful of children. Like a pair of tag team wrestlers, we had faced a fearsome opponent and learned that we could count on each other in a crisis.

In the cool, clear weather left by the front, we sailed back up the island chain and crossed the Gulf Stream to Florida. After a short stop in the Keys to rest, we returned to Twin Dolphin Marina on the west coast of Florida, where we were welcomed by old friends. Tying the dock lines took on metaphorical significance; the physical reconnection mirrored the emotional one as we were folded back into a community. We were happy we had gotten a taste of the traveling life but also felt that we had made the right decision in coming back.

We spent the next eight weeks working on boat projects and preparing for the birth of our baby. We hired a carpenter to build a new and stronger catwalk and new hatches over the fuel tank fill ports. On the interior, he built a custom crib in what would be the fifth cabin in the port hull. He built a new cockpit table, a new salon table, new countertops for the galley, and a new chart table which allowed for the installation of a washing machine. (We balked at the idea of washing all those cloth diapers by hand!) We hired a professional to make new kid- and pet-friendly cushions for the salon—double covered with an impervious vinyl underneath and a Sunbrella outer layer that fastened with Velcro so it could be easily removed and washed.

The boat's interior transformation was almost as dramatic as that of my body. As the weeks wore on and my belly burgeoned, our friends at the marina would stop me and ask, "When are you going to have that baby?!" Or exclaim, "You look ready to pop!" Indeed, as Sarah's seventh birthday approached, I grew more and more uncomfortable. Climbing into our bunk (several times each night as the baby put pressure on my bladder) became unsafe, so I slept upstairs on the settees. By the due date, the nursery was ready, clothes and diapers had been purchased, washed, and folded tidily in their cubbies, and the only thing missing was the new little person. During my regular visits with the midwife at a birth center in Sarasota, nothing seemed out of the ordinary, and all we could do was wait.

As I grew larger and more unwieldy, the children were asked to take on more responsibilities. Sam, at four, was now able to take on a few tasks, so we revamped the chore chart, finding ways to divide labor so that when the baby arrived, things would still get done in an orderly manner. Even Jay chipped in when he wasn't working or traveling for business. He came up with a strategy that we came to call "team shop." He would take the four kids to the store with my shopping list and send Eli and Aaron, now aged ten and nine, respectively, off on missions to fill handheld baskets with various items. Meanwhile, he would contain four-year-old Sam in the cart and ask Sarah to show him what products and brands we normally bought. They would meet at checkout, bag the groceries, and be home in record time. After the baby's arrival, I would be doubly grateful for the "team shop" training Jay had given the kids; teamwork is a gift that keeps on giving.

We had been in nesting mode for more than two months,

when, one afternoon, as I sewed the shower curtain for the newly renovated head in our cabin, my water broke. The baby, whose arrival we expected in the last half of April, was ten days overdue. We were more than ready. I rang the bell and all hands came running. I told everyone to pack their overnight bags—it was time! Jay walked the boys to the neighbor's boat and called his mom. They would go to Clearwater to be with their Mimi and Pappy while Jay took me to the birth center in Sarasota, twenty minutes away. Sarah had decided she wanted to be present at the birth, so she climbed into the back of our minivan, next to the empty infant car seat, and we were off. A labor that should have taken hours, (it was our fifth child, for heaven's sake!) lasted a day and a half, and Rachel was born on May second, just hours before Sarah turned seven. We sat on the bed at midnight with a sleeping newborn between us, and Jay snapped a picture as she blew out the candles on a chocolate cake the midwife had brought in on a tray. God's timing could not have been more perfect: Sarah had prayed for a baby sister, and one arrived just in time to be her birthday present.

After a few days' recovery at my in-laws quiet, comfortable home, I returned to the boat with Rachel in my arms. I laid her in the Moses basket we had bought at the George Town straw market in the Bahamas. I looked around my floating home. As usual, there were projects still in progress, and with Jay working and taking care of the kids simultaneously, the cabin looked like ground zero after an explosion of toys. But it was *home* and we were now a family of *seven*. For a while, we would have the help of friends at the marina and their generous meal deliveries, but even with their loving assistance, adding a whole new person was a lot of work. We relaxed our expectations for the first few weeks while I rested and fell into the rhythm of caring for a newborn whose daily itinerary read *eat-poop-sleep-repeat*, and relied on the family to get household tasks done.

Rachel was only a few days old when we sent Eli and Aaron out on a new team shopping mission. It was a Sunday, Mother's Day, in fact, and we had given them a twenty-dollar bill and told them to walk to the Farmer's Market to buy fresh-baked bread and local strawberries for breakfast. The marina was located on the waterfront of a small town which closed the main street to traffic each Sunday and held a community-supported market with produce stands, artisans, bakers, and live music. It was only a few blocks away, and though they looked a little tentative as they headed up the dock, we knew our boys could handle the task. I had taken them to the market often, and they knew how to safely cross a street. They returned half an hour later with a crusty Italian loaf in a paper bag and two pints of strawberries—mission accomplished! When I asked for the change, they looked at me sheepishly.

"Well…" they said, shuffling their feet.

Eli quickly explained, "We had seven dollars left after buying the bread and strawberries, but then on the way back we saw this booth selling necklaces made from abalone shells. We found a turtle necklace we thought Sarah would like. It was eight dollars, but we told the guy that all we had was seven and he sold it to us. I know it wasn't our money to spend…Are you mad?"

My boys had haggled with a salesman to buy their sister a gift. Was I *mad*? I was incredulous, touched, and overjoyed. I couldn't imagine a better Mother's Day gift. I beamed and offered my blessing. Spending 24/7/365 with our kids sometimes made us feel like all we did was correct bad behavior and resolve conflicts. To know that they could accomplish a shared goal and go above and beyond by doing something thoughtful for their sister was amazing.

. . .

While teamwork and good attitudes are hit and miss for chore time, emergencies are where we shine. People who can't seem to work together to get a meal on the table on a normal day suddenly pull together when there's a sense of urgency. Docking in the wind, sudden thunderstorms, dragging anchor, and man-overboard drills require all hands on deck. Thankfully, we've never had a true emergency at sea, and "man overboard" usually means "man off the dock." Everyone in our family has fallen "in the drink," either stepping on or off a dock, climbing in or out of the dinghy, or, comically, tumbling from the hammock that swings from the arch between the transoms. Even the captain has fallen in—though not since he was a child on his parents' boat. These stories all have happy endings, with the worst being barnacle scrapes and cold, salty swims. But everyone understands the danger of getting hurt or drowning, and we take overboard drills very seriously.

When Rachel was about two years old, we were walking down a dock at our marina in Fort Pierce, Florida one day, heading to the public library. Distracted by a steam-cleaning operation going on in a nearby mega yacht, our toe-headed toddler got too close to the edge and slipped off the dock. I immediately dropped my bag and jumped in after her and shouted up at Aaron, Sarah, and Sam, "Go get Dad! And grab some flotation...and hurry!" They ran back down the dock to get help.

The dock was fixed, about six feet above us—and there was no safety ladder nearby. Luckily, it was slack tide. I treaded water and kept Rachel calm. She had remembered her survival swimming skills and didn't freak out. Aaron and Sarah ran back to tell me that Dad was on the way. In short order, Jay zoomed up in the dinghy and started to hand me a floating cushion. I laughed and said, "Why don't you just take Rachel and I'll climb in the boat?" Eli was in the cockpit, having missed the initial emergency, ready with a towel to receive his cold, wet sister.

After we got showered and dried off, we had our habitual debriefing: *What went right? What went wrong? And how could we improve our responses if it happened again?* We had all worked together to get Rachel and me out of the water in a timely manner. But in our rush, we made all sorts of adrenaline-fueled, silly mistakes. If I had looked around, I could have swum under the dock to a boat on the other side and simply climbed out on the stern. The kids could have grabbed floating cushions or a life preserver from a nearby boat (they didn't think of this because there's an unwritten rule that you never step on someone's boat without express permission). Jay didn't need to bring floatation. And Eli, who had been in the bathroom when the bell rang, could have put on his pants before he ran to help!

Though we have made some laughable mistakes during emergency maneuvers, what we learned from them is that we could count on each other. If any one of us were in trouble, the others would be ready to offer aid and assistance. People on a team must be able to trust each other, encourage each other, act for the greater good, and compensate for the others' mistakes. For a team to gel, they have to know each other's strengths and weaknesses, to problem solve and listen to others' ideas. They must work toward a common goal and be able to compromise. They must have the courage to try again after failure. For whatever reason, hardship can be a catalyst for this kind of cooperation.

. . .

My favorite teamwork story took place after the worst passage we ever had on *Take Two*, nearly ten years after buying the boat, while we were cruising in the Western Caribbean. The only redeeming quality of that fast and furious trip between Panama and San Andrés, Isla de Colombia, off the coast of Nicaragua, was that it was over quickly. We left Portobelo using the first possible weather window since the arrival of the "Christmas Winds"—stronger-than-normal trade winds, which kick up daily gales and nasty seas off the coast of Colombia in the winter time. (These gales are what send the Bocas del Toro area enough swell to make it a rare Caribbean surfing destination.) We had stayed longer than we planned; it was the end of January and our cruising permit was about to expire. We had stocked up in Colón and Panama City and spent many happy hours with friends from Texas and Turkey on a spacious gulet named *Jubilee*. We knew it was time to go, so we took our leave and sailed on a weather window that seemed "good enough," but which turned out to be wishful thinking.

That trip was constant slap and bang, waves hitting us from different directions, some peaking beneath the bridge deck with loud booms. Every available settee was taken up by crew members trying (unsuccessfully) to sleep away the miserable hours. Things were falling down and crashing in distant corners of the boat. We could barely move, let alone eat, drink, or go below. Upon arrival, we found that seawater had forced itself into places that normally stay dry on passage; everything was damp or crusty with salt. We had traveled an unprecedented 240 nautical miles in thirty hours —with sails reefed.

Despite the speed and safe arrival, we were demoralized. How could we still be making the same mistakes after all our years of sailing? Anchor down and drinks in hand, Jay and I apologized to our crew for the unexpectedly rough trip and praised them for their toughness and patience. Though we felt like trash, and the boat looked like the aftermath of a natural disaster, Jay and I needed to go ashore to meet our agent to get the customs and immigration paperwork started. We lowered the dinghy, hopped in with our bag full of passports and official documents, and headed to shore.

Check-in took a while, and when we finally got back to the boat, the sight that met our eyes was unforgettable: everything was clean, tidy, organized, and, in a word, shipshape. Aaron had tidied up the interior, stowing loose items and collecting salty or damp clothes. Eli had hosed down the cockpit. Sarah had taken care of all the dishes, cleaned the galley, and started a load of laundry. Sam and Rachel had swept and mopped the salty floors. I was speechless. This was my crew, a collection of mismatched personalities, working together without being told! They had acted unselfishly, responsibly, and cooperatively—a parent's dream, a sign of success, confirmation that *we are doing something right*. I took a mental snapshot of that moment, so that when things are not going well, I will remember what's possible and take courage. And then I took them out for ice cream.

From these success stories one might get the wrong impression. The truth is that we more often resemble the cast of *Gilligan's Island* than an Olympic sailing team. So often, we fail in the teamwork department. I am guilty of poor leadership—failing to communicate clearly, expecting too much or too little, micromanaging, and becoming irrational when things don't go my way. Jay is quiet by nature and leads by example instead of explanation. But when things get unpleasant, he tends to withdraw. The children argue like a team of lawyers, causing me to throw up my hands and repeat my father's lament, "Why can't you all just get along?" There is nothing special about us; we are a sarcastic, argumentative, and stubborn bunch of people. But our lifestyle has presented challenges that are too hard for any one of us to manage alone and which force us to cooperate.

When we arrive at a marina, for example, everyone must know his role and execute it well to prevent accident or injury. We signal *all hands on deck* and Jay gently guides the boat into a slip while I prepare to throw a bow line to a dock hand. One kid will be on the stern with a fender in hand, and another on the bow. A third gets ready to toss a spring line from amidships. A fourth waits for the signal to hop on the dock and help secure lines. A fifth stands by to do whatever else might be needed. This coordination is the culmination of years of practice, a multitude of mistakes, and the usual sacrifice of blood, sweat, and tears. While I'm not saying that living on a boat is the only way to turn a group of people into a team, it is a *good* way.

# 9

## BATTEN DOWN THE HATCHES
### HARDSHIP AND HOPE

*June 2012. It has just begun to drizzle, and it is late in the day. Nothing has gone smoothly. We are getting ready to leave the marina after a year of living on the dock in Bradenton, Florida. The engines are running, and we feel a bit harried. The older kids, excited about the departure but tired of waiting for us to finalize preparations, are playing on the dock. Aaron slips and falls and comes in with a bloody knee. Jay and I look at each other. We're wearing rain jackets, small children are crying, we're losing daylight, and it all feels so...wrong. We turn off the engines and decide to wait a day. The weather is deteriorating—it's June and we are facing the first wave of tropical weather. We were hoping to use the easterly breeze before the arrival of a storm to work our way down the coast to the protected waters of Charlotte Harbor and Pelican Bay, near Cayo Costa State Park, a place we know well. But leaving is hard, and the weather forecast does not always show us what the conditions are really like.*

*We wake the next morning to sunshine. The day spent preparing means that we are now rested and ready to go. This has become a pattern for us—it will often take two tries to leave—the day we think we are leaving, usually spent doing last-minute chores, running errands, or making an unexpected repair, and the day we actually leave. It gives a whole new meaning to our boat name.* We pull away from the dock, waving goodbye to new friends we have made and old friends to whom we have grown closer.

We sail out of the Manatee River and into Tampa Bay. By the time we get out into the Gulf of Mexico, we are beating into the wind and facing an increasingly choppy sea, and we do not see the east wind that will help us go south overnight to Boca Grande. We have been sitting in one place for a whole year; between adjusting to a new baby and renovating the interior of the boat, we rarely untied the dock lines. This is not the time to bravely battle rough seas, but we cannot bear to turn around and go back to the dock after we have worked so hard to leave. While securing loose items, closing deck hatches, and scrambling to prepare the boat for rough seas, we try to think of a place nearby where we could go to anchor and ride out the storm.

Somebody suggests a place we used to anchor on weekends when we first bought the boat: Terra Ceia Bay. Just north of the Manatee River, this quiet mangrove bay is just the right kind of place to seek shelter. We turn the boat around and suddenly, instead of forcing our way upwind, we are enjoying a gorgeous downwind run in pleasant conditions. We enter the bay early enough in the day that we have time to explore, find a nice anchorage, let the kids swim, go kayaking, and remember what it's like to be a boat and not a "dockominium." We anchor near aptly named Bird Island, and dozens of beady eyes glint with suspicion as we wet our new Manson eighty-pound anchor for the first time. The anchor is huge and has a roll bar for easy resetting during a shift in wind or tide. Unbeknownst to us, this will be its first real challenge—trial by tropical storm.

We have a saying about weather forecasts—their purpose is to give us the confidence to leave, but once we get out there, we'll deal with whatever we find. Sailors spend a lot of time studying weather patterns and predictions, and planning routes and departure dates accordingly, but they also recognize the risk of finding conditions worse than they expected and put plans in place for rough seas, whether it's turning around, finding a bailout point, reefing the sails, securing loose items, or literally battening the hatches, sealing out rain or salt spray. Metaphorically, of course, we all *batten down the hatches* when facing the proverbial storms in life. Just as sailors study weather forecasts and prepare their boats before weighing anchor, landlubbers also prepare for uncertain times by reading the news, following stock market trends, buying insurance, taking proactive medical exams, stocking cupboards, and putting money in savings. In short, people often hope for the best but prepare for the worst.

Despite it's being the official beginning of hurricane season, we are not afraid to travel in June, or even July, for that matter. Though we might get a little nervous in August or September, we have often lived whole years inside the "hurricane box" proscribed by insurance companies. Simply put, we are Florida people, and we know what a hurricane is, and what it isn't. The year we moved back to the Sunshine State, we experienced the effects of four hurricanes—Dean, Francis, Ivan, and Jean. We had boarded up the windows in our house for two of the storms and left town for another. After the last one, we called the window company and paid the big bucks to put in impact-resistant glass. Of course, in accordance with the umbrella theory, not a single hurricane threatened to hit Clearwater again as long as we owned that house.

Then we bought a boat. Moving aboard taught us what it meant to live close to nature: knowing the phases of the moon, sensing the shifts in wind and tide, and recognizing the diurnal patterns of sea birds. But it also meant preparing for approaching storms and being vulnerable to ever-changing weather conditions and temperatures. It filled us with wonder, but also made us feel small and exposed. Even the summer thunderstorms we weathered at the dock were exciting—sometimes forty-knot gusts would push the hull up against the dock and make the rigging scream. Rain pelting the decks made a deafening sound in the interior. The first summer we owned *Take Two*, Tampa Bay had a hurricane threat and we had to move our boat from a dock end to a slip on the seawall, where we practiced spider-webbing ourselves in. Thankfully, it was a false alarm, but we began to realize what it meant to have a floating home in the path of destruction.

During our first real storm at sea on the Bahama Banks in 2011, we gained insight about our boat and ourselves, as well as the power of Mother Nature. We noted what *not* to do when a cold front approaches, and we picked up survival strategies. Our admiration for the sailors of old—the ones who used barometers and observation of the elements to predict the weather—grew enormously, even as we checked radar, listened to weather broadcasts over the radio, and interpreted data we received by email or on the internet. We became careful and tried to avoid passage conditions that would cause discomfort and distress for our young crew. We would no longer sail according to dates on the calendar, but rather when all the stars aligned and we felt ready.

And that summer storm we weathered in Terra Ceia Bay in 2012? It became Tropical Storm Debby, which hovered for a week, causing extensive flooding and wind damage. We float, so the flooding wasn't really an issue, but wind and rain can create conditions that make for a harrowing experience on a boat with five small children. Even in our little protected bay, the wind created waves that made it feel like we were on an ocean passage upwind with a constant pitching motion. Rachel was just learning to walk, and she thought it was great fun. She would run down the sloping floor of the boat, wait for the next wave, and then run down the slope in the other direction. When she tired of this, we put her in the swing in the cockpit, and the rocking of the boat did the rest, no batteries required.

But entertaining a one-year-old wasn't the only problem. The other kids got cabin fever too. Between rainstorms, ten-year-old Eli, nine-year-old Aaron, eight-year-old Sarah, and four-year-old Sam would don windbreakers and life vests and go out on the trampolines for fresh air, tumbling, and wrestling out their wiggles. When the rain started up again, droplets blown at thirty-five knots felt like driving needles and forced everyone indoors. I popped popcorn and declared a matinee movie marathon: we watched a *Star Wars* movie every day at 2:00 (during Rachel's nap)—hoping that we would run out of storm before we ran out of sequels.

We had plenty of food, having prepared for a passage around the Keys to the east coast of Florida. With a generator and watermaker, we were in some ways better off than the landlubbers who lost power and water. Though we felt isolated, we knew we were anchored in a safe place. But it turned out that we were neither isolated, nor considered safe. One night, a bright light appeared on shore—a powerful flashlight turning its beam on *Take Two*. Someone on land directly behind the boat was keeping an eye on us. The next morning, using binoculars, he read our boat name and looked us up on the internet. He read our blog and contacted us by email. We were a few hundred feet from shore, and he was afraid that we were going to drag onto his property. He initially thought we were just weekenders caught unawares, but after reading our blog, he realized that we were probably okay. He jokingly offered to "float us a pizza" if we needed it. We exchanged messages, reassuring him that our anchor was holding, and that we were safe, if a bit uncomfortable. We got friendly, and when the storm ended a few days later, he kindly invited us ashore to let the kids run around.

Steve and his wife had owned a sailboat and had traveled the world before settling down on a large tract of land bordering this quiet bay to start an organic farm, applying the self-sufficiency learned at sea to their land life. We spent a lovely day at his property, helping him pick up tree branches broken in the storm, letting the kids run free and climb trees, and even borrowing his canoe to explore the shoreline. We felt as if we had made a passage and arrived in another country—even meeting another traveler and hearing his stories. When it came time to leave again, we were ready. We knew we could handle the sea conditions after five days of being storm tossed, and more importantly, we remembered one of the things we liked about living on a boat: the surprise encounters with new people in new places. It motivated us to keep going.

We had endured a tropical storm on our boat. Surviving it made us stronger and built trust in our seaworthy vessel and sturdy anchor. Later that same year, we had another opportunity to see rough conditions, this time at a dock. When a late-season hurricane, Superstorm Sandy, raked up the East Coast, we were at a marina in Fort Pierce, Florida. Ahead of the storm, we took exploratory dinghy rides all over the surrounding area, looking for a protected patch of mangroves in which to tie our boat—planning for the worst. We watched the forecast and as conditions changed, we decided that there was no reason to leave the dock. We might see gale-force winds, but nothing more serious. We doubled dock lines and added more fenders to cushion the hull against the pilings.

As the storm passed, the wind in the rigging howled and screamed and the dock lines and fenders creaked and groaned as they held the boat on and off the dock, respectively. The motion became intense, and we decided to get off the boat. Jay stepped carefully from the moving deck of *Take Two* to the wet surface of the wooden dock and I passed each kid over to him, holding them by the hand until their footing was secure. I carried Rachel as we walked down the dock in the strong wind to the van. We drove to the public park on South Hutchinson Island that borders the inlet, where we could observe—from a distance—the conditions whipped up by the storm. The sea was in a fury, it's disturbed surface white and frothy. Waves were crashing over the boulders that form the jetty. There were geysers, walls of spray, and banks of foam showering the rocks, which were normally high and dry. Rachel was in my baby backpack and was wriggling and trying to get down as she watched the older kids laugh and shriek, leaning into a wind so strong it could prop them up. There was no way I would set her down—she was so little I feared she might blow away! Two manatees were sheltering behind the breakwater, and we could see them sloshing around and coming up periodically for air. I was so grateful to have my feet on firm ground and not on the deck of a pitching boat.

Experiencing those big storms taught us to develop protocols and contingency plans. Tropical storms are forecasted long in advance, and we usually have a chance to prepare ahead of time and decide the best course of action. We always ask the same questions: *What is the likelihood that we will sustain a direct hit? How much wind will we get? Where do we want to be for this storm? How do we secure the boat and keep the crew safe?*

We've learned what must be done ahead of time: moving the

boat or doubling dock lines, clearing the decks of anything that can blow away, and wrapping the jib tightly so it can't catch a gust and unroll, violently flogging itself to shreds and possibly dismasting the boat. We tie down the main, remove awnings or other things that create windage, and secure kayaks and solar panels. We buy extra supplies, wash laundry, and run last-minute errands—just like we do when we're getting ready for a passage. We make storm plans that account for best and worst case scenarios. Because we are mobile, we could get out of the path of severe weather if we need to, though this plan requires making the decision to take evasive action long before the need arises. Also, NOAA's hurricane forecasts sometimes put the entire state of Florida inside the cone of uncertainty, which leaves us feeling confused about where we should go to be safe. We try to take the most conservative option and have been very fortunate not to have damaged or lost our boat.

Storms at sea are another story entirely. We have plans and protocols there, too, but we have to be more flexible and quicker on our feet because we don't always have advance warning. In the summer of 2014, for example, we experienced a massive thunderstorm off Cape Canaveral that I will never forget. We had been listening to the radio because there was supposed to be a rocket launch that day, and we had to be a certain number of miles offshore to be out of the fallout zone. We were excited to have front row seats on the water—a unique vantage point for a launch. And then we heard the disappointing words, "mission scrubbed due to weather."

"What weather?" we thought. It was a breezy, sunny July afternoon. We were sailing along enjoying pleasant conditions. We didn't see any threatening black skies…yet. But those afternoon squalls can arrive like a high-speed train and pack a powerful punch. And no matter what you do to prepare—furl the sails and batten the hatches—things can get chaotic very quickly. Fluffy white clouds materialized in the distance. The clouds darkened and took on the look of muscular bullies in a back alley, rolling towards us at a surprising speed. The sea turned steely gray, the threatening cumulonimbus blocking out the sun. A cold blast of air came over the water. And then it was upon us: blinding rain, lightning, continuous rumbles of thunder, and confused seas. All we could do was turn downwind and run with the storm, bouncing up and down in the waves. I stood in the cockpit, wind whipping my hair around my face, flashes of pink and blue lighting up the surface of the ocean, creating a strobe effect on the falling rain and ocean spray. It was both terrifying and beautiful —I've never felt so alive, and so grateful when it was over. Two cruise ships were coming out of port that afternoon. We knew they were there; we were receiving their radar reflections and AIS signals, but even at a too-close one nautical mile, visibility was so bad that we never saw them.

That was the beginning of an offshore sailing trip up the Atlantic seaboard from Florida to Virginia that continued to stretch us, each leg presenting unique situations and challenges. From the shifting sandbars and heavy current of the St. Augustine inlet to the commercial port of Charleston where we were overshadowed by passing container ships, to the choppy waters off of Diamond Shoals as we sailed past the Outer Banks, we added new skills to our wheelhouse. Once we were in the Chesapeake Bay, a huge body of inland water, we discovered that it had its own conditions and moods. We arrived in the dark, of course, motoring up the York River to anchor blindly, waking in the morning to see that we were a stone's throw from the Yorktown Battlefield, the scene of the final battles of the Revolutionary War. We spent the rest of the summer sailing from one anchorage to another, exploring rivers and creeks and bays, homeschooling in history parks, colonial towns, nature preserves, and forts as we made our way up the Potomac towards Washington D.C. During our three weeks at the Capitol Yacht Club, we visited nearly every monument on the National Mall and every museum of the Smithsonian Institution, all walking distance from the marina.

One chilly October morning when the Potomac River dumped us back into the Chesapeake, we experienced what some have called the "Chesapeake Chop," short, steep, multidirectional waves that leave the hardiest of us feeling queasy. I had made a pot of oatmeal that morning which no one ate. We began to beat our way toward Hampton, Virginia, a long, hard day in cold spray under gray skies. It was not our idea of pleasure boating.

"This is miserable. Do we need to go all the way to Hampton today?" I asked.

"I agree, it is miserable," Jay responded. "Maybe there's somewhere we could go to get out of this. It's supposed to be calming down, so maybe it'll be better tomorrow."

"I could look at the chart and see if there's somewhere we could bail out. Maybe a small river or bay? Let's go find a waterfront town somewhere…and take the kids out for lunch! After this, everyone's going to be hungry. I could really go for a cheeseburger and fries." Small ears perked up.

Jay threw me a sidelong glance. "Careful there. Don't be making promises you can't keep."

But I fully intended to make amends for the missed breakfast and morning spent bashing in the waves. Jay was right, of course, optimism and its enthusiastic promises have the downside of sometimes causing disappointment. I diligently studied the chart. We knew there were lots of rivers that empty into the estuary, rivers with native sounding names like Rappahannock, Pocomoke, and Wicomico. Lesser known were the creeks of

Fleet's Bay in the Northern Neck of Virginia, like Indian Creek, Bell's Creek, and Henry's Creek. But they looked like exits off the Highway of Discomfort to us. As soon as we turned and headed toward shore, we all breathed a sigh of relief.

We chose an anchorage based on its location on the map. It was up a narrow creek within walking distance of a town: Waverly Cove near the town of Kilmarnock. Like a fairy tale ending, the wind shut off when we entered the placid backwater, and the sun came out from behind the clouds as we anchored. We got in the dinghy, found the public dock at the end of Route 608, tied up, and started walking. As if by magic, the road led us straight to the adorable main street of a tiny town, and on it, the Northern Neck Burger Company—a gourmet burger place with a fun atmosphere and amazing food. Jay just shook his head incredulously, saying, "I *cannot* believe that we pulled this off." Happy and satisfied, we walked back to the dinghy that afternoon in autumn sunshine, laughing and talking, and spent the evening listening to the chirp of summer's last crickets and the honk of geese flying south overhead. We made it to Hampton the next afternoon, where we would wait for weather to head offshore, making our own migration to warmer climes.

As winter approached in the North Atlantic, cold fronts began to sweep down from the arctic and weather windows narrowed. With a sixty-eight foot mast, going through the Intracoastal Waterway with its fixed bridges wasn't an option, so we waited for the right conditions to go out around Cape Hatteras. We had time to sightsee while we waited, even viewing the wreckage of the historic ironclad ship, the USS Monitor, at the Mariner's Museum in Newport News. Artifacts from that and other ships were in an exhibit hall featuring a large wall map of Hatteras. I pointed to the mural, and said, "Look, kids! That's where we're heading next week!" The words that curved around the shoals read, "Graveyard of the Atlantic." Eli, at thirteen, rolled his eyes, and responded sarcastically, "Sounds like fun."

We needed only a day or two of clear weather to get out around the cape, and then we could bail out in the Carolinas if conditions deteriorated. But time was of the essence. It was getting cold, with freezing temperatures at night and frost on our decks in the morning. We bundled up, slept with woolen caps on, and baked muffins every morning to heat the interior. Finally, we saw a window of opportunity: a few days of brisk wind from the north, and then a shift to the east that would make for a speedy trip south. The assumption for passage making on our catamaran is that if there is wind, we sail, but it will be fast and uncomfortable. If there is no wind, we motor, and it will be slow and boring. Fast *and* comfortable is a rare occurrence. This would be a passage of the first description.

On our departure day, Jay and I dug out our foul weather gear: waterproof pants, hooded jackets, and rubber boots that we almost never used in Florida. We put them on over our winter clothes in Hampton, not knowing that we would live in them for five days until we reached our destination in Florida. The kids made cozy nests in the salon with blankets and pillows, knowing it might be too rough to go down companionways and climb in and out of bunks. Jay took his seasickness medication and gave twelve-year-old Aaron a half dose. Sarah, age ten, took a fruit flavored chewable tablet, but Sam and Rachel, age seven and three, refused theirs, saying it made them feel "barfy" just to smell it. Eli and I, oblivious to the motion, worked through an invisible checklist to ready the boat for heavy seas: duct tape the drawers and fridge door to keep them from flying open, cover the scuppers to prevent geysers in the cockpit, make sure all hatches are closed and battened, and secure all loose items that might fall.

We raised the mainsail as we exited the Chesapeake. Turning

south, the sail filled with wind, the lines grew taut, and the wake behind the boat began to gurgle and gush. We unfurled the jib and picked up speed. With the waves behind us, we began to surf down the leading edge of the slick faces topped with curls of foam. The motion more closely resembled a bucking bronco than a sleigh ride, as the waves picked up the boat and slid under our bows, dropping us into the mounting swell behind it. Jay set our course and turned on the autopilot, then wedged himself into a corner of the cockpit where he could keep a close eye on the instruments. I checked to see if the kids needed anything, then found my own cozy corner and cuddled up with Rachel on my lap.

That evening, close to sunset, we heard the U.S. Navy on the VHF making an official sounding announcement. The first time we heard it, all we caught was "…practice with live ammunition…" We turned up the volume and listened more closely as they repeated the message. They were warning vessels in the vicinity to steer a course around a box with corners marked by GPS coordinates, then rattled off a bunch of numbers. A few minutes later, this time with paper and pen in hand, we wrote down the coordinates, matched them to the chart, and realized that *we were inside their box!* We adjusted the sails and turned ninety degrees to avoid their live-ammunition practice. The motion of the waves now fell directly on our beam, and we began to rock side to side. The bell swung wildly and started ringing with each wave that passed under our keels. We took it down and stowed it somewhere safe. For a few hours, until we could get back on course, we rode the roller coaster of the Atlantic Ocean, watching the red sparks of live ammo in the distance as the Navy ship carried on with its maneuvers.

The next morning, I was on watch as the sky turned pale blue and the first rays of light broke the surface of the horizon. The waves rose up behind me where I sat in the captain's chair; as we slid from peak to trough, our boat dipped below their frothy crests. As the buoyant hulls lifted us over the wavetops, I could see what looked like an endless green mountain range, and from the valleys, the morning light shone green through their translucent surfaces. Suddenly, I noticed a dark shadow in the bottle-green light of the wave behind us. A silvery fin broke the surface. Without warning, a dolphin sprang out of the wave with a splash, slid under the stern of our boat and disappeared. A moment later, the same silhouette appeared in the mounting emerald swell behind us. I watched, transfixed: the dolphin was *riding* the underside of the waves like an inverted surfer. I was afraid to break the spell but wanted someone else to see this phenomenon. Without turning away, I called to the kids, but got no response from the lumps under the blankets. After a few minutes, the dolphin vanished with a flick of its tail, and I was alone in the cockpit, left with a sense of wonder, changing my anxiety into irresistible joy.

For some people, all these unpredictable weather conditions and constantly changing plans would be enough reason to hang up the boat hook and go back to living ashore. Why do we go back out when we've had a bad day? Why do we keep sailing year after year after year, with a large family to boot? Why are we undaunted? Perhaps we are gluttons for punishment, or maybe we just like to do things the hard way. But we also feel that the hardships are worth it. For us, there have been short-term losses but long-term gains, and higher mountains to balance the lower valleys.

The storm at sea makes a particularly poignant metaphor— even the landlubber can appreciate the way overcoming one challenge can prepare a person for another. Who hasn't felt storm tossed, all at sea, as vulnerable as a boat on the ocean? Struggle and hardship are universal: every danger we dodge, every storm we survive, every difficulty we surmount, only strengthens our resolve to continue living, to see what adventures we can find, and to develop our inner strength. A storm at sea, or in life, leaves us vulnerable, but also grows us in ways we might not otherwise discover. Sometimes I have felt confident, working on deck in the middle of wild weather, full of gratitude and amazement despite awareness of the danger. Other times I have wedged myself into a corner and cried. But even then, humbled completely, reduced to asking God to calm the storm or give me the strength to withstand it, I have discovered traits like Hope and Optimism, Perseverance, and a Stubborn Will to Survive. And each success prepares me for the next test. All the hardship serves a purpose.

When we face obstacles, like storms at sea, we also learn to accept circumstances that we can't control, like the weather, or the moods of the ocean. We are forced to adapt. We toughen up, we learn to deal with discomfort and keep our spirits up in the midst of difficulty. Not every human needs to experience the fury of Mother Ocean to learn these lessons, but she is an expedient teacher. And while I hate how fickle she can be—calm one minute and stormy the next—I love the exhilaration of a fast sail, the glory of a sunrise at sea, the joy of dolphins playing, and the peace of an anchorage at the end of the day. Even as I batten down the hatches, I know that storms don't last forever, and I'm willing to put up with the ocean's moods to experience its wild beauty. That's the tradeoff with an adventurous life—you can't have one without the other.

# 10

## GETTING SHIPSHAPE
### ORGANIZED CHAOS

*November 2014. The spicy smell of pumpkin pie and apple crisp seeps out of my warm oven. A breeze blows through my galley, the weather in Florida finally showing signs of cooling. When the dessert comes out, I'll use the residual heat to warm up the sweet potato casserole and dinner rolls I made earlier. For a moment before everyone arrives, the boat is quiet, and the afternoon sun shines in on our handmade placemats and autumn centerpiece.*

*The salon is as clean and organized as it ever gets. Despite the fact that each drawer, box, bin, and shelf has a nice little printed label, the contents of these carefully compartmentalized spaces are frequently elsewhere. "A place for everything," sure, but "everything in its place?" Not so much. But just for today, the odds and ends have disappeared into lockers, nooks and crannies, and our cozy salon looks like a girl ready for her first date.*

*Thanksgiving is my favorite holiday. What's not to like about food and family and gratitude? Yes, it can get a little crazy. Yes, I spend days getting ready. Yes, we cook and make a big mess and invite a lot of people into our small space. But no, there is no such thing as too much love, too much friendship, or too much pie. Our friend Ben uses* the phrase "organized chaos" to describe Thanksgiving on Take Two. With six adults and seven children this year, there certainly will be a lot of noise and movement, but then, the kids are helpful, and many hands make light work, right?

Everyone has a job. The kids have already helped clean the boat and set the table. Jay has taken them over to Ben's property for the afternoon to make a bonfire and conduct a deep-fry turkey experiment while I finish the baking. Ben has the kind of place where one can risk cooking a big bird in hot oil over an open flame. They come home smelling like smoke, hot turkey in hand.

Ben's friend Carla comes with stuffing—mountains of stuffing. Another sailing family, Marina and Joseph and their two boys, Owen and Zuber from Little Wing, have joined us, bearing gifts of steamed green beans and mashed potatoes. Just finding a place to put all the food is a problem. Even though we have two tables and a huge galley (for a boat), that many people carrying dishes makes our cabin feel automatically chaotic. We get it all sorted out and drinks poured (and spilled, and poured again), bless the meal and settle in for a nice dinner.

Because we have seating for sixteen to twenty people, we often end up hosting holidays for other boaters and last minute acquaintances. We may not have a house, but we are not homeless, and one more is always welcome. By dessert, friends from the marina have wandered down the dock and joined us. We let the kids go first, and when they disappear down below into cabins to play games or build with Lego, the grown-ups linger over coffee or a glass of wine and dessert. Peace settles over the salon as the pie disappears, and I finally sit still and bask in the holiday afterglow.

This is how much I love Thanksgiving: we're doing it all again tomorrow. We've invited people over again for Thanksgiving Part II, where I will ladle my Thanksgiving Leftovers Soup with stuffing *dumplings out of a twelve-quart stock pot. And for dessert? More pie, of course.*

---

"Shipshape and Bristol fashion" is an expression that grew out of the busy west coast seaport of Bristol, England centuries ago and it has come to mean "everything in order and ready to go." It carries the connotation of organization, orderliness, and preparation. For us, shipshape means that the boat is kept in good working order, that it can be quickly tidied, restocked, and prepared to set sail should the need or whim arise suddenly. While we hold this as the ideal, the reality looks a little messier. Sometimes we are in the middle of a major refit, or Jay is working on a project that requires the use of nearly every tool he owns and the rearranging of our interior so that it resembles the aftermath of a tornado. For an old wooden boat, repair and upkeep are constant and ongoing, and sometimes we are forced to stay in port when we would rather travel.

We are the fourth owners of *Take Two*, so there are some maintenance mysteries to solve and mistakes to remedy. Also, because she is custom built completely of wood, she's flexible, buoyant, long-lasting, and strong, but also susceptible to dry-rot—a fungus that grows when wood is exposed to air and fresh water, like if there is a crack in the paint, or a screw has not been properly bedded in epoxy. It manifests as soft spots or places where the surface is cracking, flaking, or chipping. I have even been alarmed to see actual mushrooms growing on *Take Two*, like the ones you see on a rotting tree in an old growth forest. Though her hull is sound, rot presents an unending maintenance issue, as we chase it down, dig it out, fill holes with epoxy, replace soft spots with new wood, and give her fresh coats of protective paint.

In addition to the material needing upkeep, a sailboat requires maintenance in multiple categories—from marine diesel engines, electrical systems, and plumbing, to sails, lines, and rigging. Not to mention upholstery, flooring, galley appliances, and other interior work. Needless to say, there is always a project list. If we had thought that buying a boat meant doing a once-and-done refit and sailing away into the sunset, we would have been sorely disappointed. Even new boats malfunction, and one definition of cruising is *fixing your boat in exotic locations*. In that case, shipshape can mean that the basic systems are functional, and the boat is good enough to get there. Tools and materials are stowed for passage and upon arrival in the next port, they are dug out again and the projects continue.

After the first few years of living aboard, our life settled into a kind of normal flux, a regular routine emerging despite changing locations. We began to take the long view: this was not a camping trip, not an experiment, but a way of life, one which we liked and wanted to continue. We learned how to split our time between traveling, fixing the boat, and working for money with which to travel and fix the boat. If we were in a season of working, or if Jay had a client who needed him on-site, we would pull into a marina, where we could be near an airport, and have easy access to amenities on shore and the benefits of community. This gave us the opportunity to get to know a place well, meet other traveling families, and experience local culture, nature, and history. While docked, we took advantage of marine chandleries and easy shipping to buy boat parts and work on projects.

As we began to upgrade the boat, we soon realized that the project list was circular—if one owns a boat long enough, he will have to fix or replace the same things more than once. When we weren't working for money or fixing the boat, we were trying to enjoy the sailing life, pursuing the elusive green flash as the sun set over an unclouded ocean horizon, or islands in paradise surrounded by warm seas filled with tropical fish.

On rare occasions, we figured out how to do these things simultaneously, like the year we hauled out in Spanish Wells in the Bahamas. With good internet and a short-term rental house for the family, Jay could work while the boat bottom was getting painted. When the boat went back in the water, we were already in the beautiful cruising grounds of Eleuthera. But usually, we're shifting between modes and trying to balance the have-to list with the want-to list.

After leaving Tampa Bay in 2012, we made our way to the east coast of Florida, to a marina in Fort Pierce with a boatyard that could haul our twenty-six-foot-wide boat, which is often a challenge. We passed up Fort Lauderdale, even though it has many yards that can handle our beam because it didn't seem family friendly. It was time for a major bottom job. A boat sitting in salt water grows a reef on its underwater surfaces, covered with creatures including, but not limited to, barnacles, worms, sponges, algae, and miniature resident crustaceans. How fast this happens depends upon the quality of the anti-fouling paint, how often the bottom is cleaned, and where and how long the boat sits. Suffice it to say that a major cleaning, sanding, and painting job needs to happen every two to three years. When *Take Two* gets hauled out for her biennial makeover, the family packs up and rents a house on land or goes on a road trip to visit family and friends.

Though the bottom of *Take Two* had been repainted in 2009, we realized that what she really needed was a fresh start—many years' worth of accumulated paint needed to be stripped off and new primer and paint applied. We pulled into Harbortown

Marina on the fourth of July—surprised to find we had front-row seats for the Fort Pierce fireworks display. We began planning the haul out, the work, and the logistics of moving the family off the boat for a couple of months. With Jay working full-time, we needed to hire someone we could trust to do the majority of the work—grinding, sanding, and painting...but who? Then we met Ben.

The story goes that I was taking the kids somewhere one day and Ben, whose wooden, hunter-green sloop, *Irony*, was docked behind us, noticed the new people climbing off the big blue catamaran—first one, then another, then another, and, unbelievably, another small child clambered over the lifelines and onto the fixed wooden dock, lining up like ducklings. And then, to his increasing amazement, Mama Duck came last with a baby strapped in a backpack. He introduced himself after the initial shock wore off, and we got to know each other over the next few weeks. Ben was the local boat guru, someone who can fix just about anything that floats. From engines to epoxy, he had years of experience and came highly recommended. He and Jay had a few conversations and Ben ended up doing the work on *Take Two*. Over the course of that project, he earned not only our trust, but also our friendship.

When the bottom was sanded, primed, and painted, and the last coat of Toreador Red went on under the bridge deck, we lowered *Take Two* back into the water and began to plan our escape from the marina. After a false start just before Thanksgiving (as in, we tried to go out the inlet, but were stopped by monstrous waves), we finally made it to the Bahamas for another season of cruising. We never seem to tire of the crystalline waters, deserted anchorages near islands with white-sand beaches, and the freedom these trips offer. Our kids call the clear waters of the Bahama Banks "God's swimming pool." For a boat family, it's a rewarding place to test their mettle—the Bahamian islands are close to Florida, but present challenges like crossing the Gulf Stream, anchoring where tides shift a hundred-and-eighty degrees, piloting in shallow water, and navigating narrow, rocky passes. We didn't realize it at the time, but Bahamas cruises were good training for challenges we would meet later.

When we returned to Fort Pierce for hurricane season 2013, we came ready to complete more big projects on Take Two. During a rough passage in the Bahamas, we had splashed our ancient Northern Lights generator, and it was no longer producing power. Our port engine, which had always run poorly, needed to be replaced, and we were suffering a bad case of might-as-well-itus—if we have to replace one or two engines, we might as well upgrade all three. The twin Volvo twenty-nine-horsepower inboard engines never provided the power we wanted. As much as we liked sailing (on a beautiful day with small sparkly waves), we realized that passage-making with kids can be tedious and slow; sometimes we wanted to shorten a long trip or time our arrival for daylight. So, we spent more time motor-sailing than we imagined, and the engines were not up to the task. After the cold front of 2011, we recognized that engines are a safety feature too. Had we possessed the horsepower to do it, we could have outrun that cold front and arrived safely in port before it hit. Would there be another time in the future when we would want more powerful and reliable engines? Probably.

So, *Take Two* went up in the slings again: nothing like seeing your home suspended from straps attached to a travelift—a kind of frame on wheels that lifts your boat out of a narrow concrete slip. Even more unsettling is watching a catamaran shake its way up tracks on a cage-like railway car, rise out of the water on a hydraulic platform, or tilt slowly up a ramp on a trailer. A boat never looks more naked or vulnerable than when it emerges, dripping, out of the water; it is akin to watching your child being rolled into surgery on a gurney. Once out of the water, the boat is blocked, literally set on stacks of large wooden blocks, or placed on jack-stands, metal supports, where it can be worked on.

We commissioned Ben to replace the three engines at once: two for propulsion and one for generating electricity. We ordered a nine-kilowatt generator and two thirty-eight horsepower marinized Kubota diesel engines from Beta Marine, and soon they were on their way from Europe. In the meantime, we had plenty of prep to do. The first step was to pull the old generator out of the boat using an A-frame gantry and heavy-duty tackle. We would also have to remove the old engines, an invasive procedure requiring us to move off the boat for the summer so we could cut hatches in the deck.

Once we had pulled that ancient beast out of the generator compartment on the foredeck, we weren't sure what to do with it. It would go to Ben's property initially—to get a closer look and see if any part of it was worth keeping (unlikely). *But if it had to be broken down,* we thought, *why not use it as a learning opportunity?* This was a chance to get a closer look at the nuts and bolts of a diesel engine. And maybe those of us who weren't mechanically inclined could use an introduction to the basics too.

We formulated a plan: Generator Summer Camp, where the kids could tear apart the old generator, label the parts, see the inner-workings, and draw diagrams. Of course, Jay and Ben were put in charge of the project (what do I know about engines?) but as an educator, I was soon involved too, making engine-part flashcards and flowcharts. We all learned something through the process, even if it was which member of our family will never be a diesel mechanic. And the flow-charts and engine diagrams were helpful for learning to trouble-shoot. I now know what could be wrong with an engine, even if I can't fix every problem. Out of the chaos of broken parts, we managed to draw meaning and purpose through learning.

By now, Ben was part of the family. We knew that *Take Two* was in safe hands when he was on the project, and we trusted him around our kids. Aaron, especially, became his little protégé, lending a helping hand and learning all he could. When we returned to Fort Pierce in the fall of 2014 after our trip up the East Coast to the Chesapeake, we decided to re-rig the boat, paint the mast, and build a hard-top for the cockpit which would provide protection from the elements and on which we could install more solar panels. We already had the aluminum framework, but the canvas that provided shade didn't keep out the rain and sea spray. A hardtop was the first step in constructing a weather-proof cockpit enclosure. Ben had finished building a large shop on his property, and it became a hive of manly activity. Jay had decided to build the hard-top himself and was constructing plywood frameworks (one concave and one convex) for the large, curved PVC board-and-fiberglass structure.

In addition to DIY boat projects, our homeschool that year included an aspect of apprenticeship and hands-on learning, as well. Once a week, the kids spent a day doing something they liked and about which they wanted to know more. Eli took flying lessons to prepare for a pilot's license; Aaron went with Ben on boat jobs as an assistant; Sarah mucked out stables in the morning and had horseback-riding lessons in the afternoon; Sam went fishing with our friend Jim whenever he had time; and Rachel (at age three-and-a-half) got to go to a mommy-and-me story time at the library and sometimes out for ice cream after. With the exception of the pilot-in-training, everyone came home dirty. It was a rich learning experience, one that would not have been available had we chosen to travel that year or enroll the kids in a traditional school. We've discovered that stationary doesn't have to mean stagnant. Sometimes when we slow down, we find hidden opportunities that will stretch and grow us as we expand our internal horizons.

Once the major boat projects were done, we were hoping to leave for a longer trip—the one we had always talked about—to the Caribbean and beyond. We had spent many years taking coastal cruises with our kids, making overnight and multi-day passages, learning how to live more simply, and forming habits and predictable routines that made living aboard easier. Even as we were getting our boat shipshape for a longer journey, we were preparing ourselves.

For actions that were repeated, like getting ready to go offshore, or arriving in a new place, we developed ordered steps that became automatic. For example, we have a list of things to do before leaving a marina, while we still have access to the dock and its unlimited power and water. We load groceries, clear the decks and clean the boat, do all the laundry, fill water and fuel tanks, and check systems like engines, steering, and batteries. When we arrive in a new place, we read the guidebook, find a good place to anchor, locate a restaurant on the water where we can dock the dinghy and have a cold beer, check in with customs and immigration if necessary, and take a walking tour of town, looking for grocery stores, laundromats, and garbage disposal facilities. Meanwhile, back on the boat, the kids can take showers and begin tidying up after the passage. Because we learned to do things the same way each time, it reduced stress and set expectations for our family. These unwritten lists and procedures provided order and comfort when everything else was changing around us. Something that once seemed chaotic and stressful became normal and orderly: controlled chaos, indeed.

At the same time, living on the boat also helped us learn to relax our expectations and go with the flow. Despite systematizing, we realized that everything takes more time and comes with a degree of uncertainty, from provisioning and laundry to making repairs and leaving for a passage. I learned to be patient and focus on completing one task each day. The kids learned that there is more to an education than schoolwork and gained valuable life skills. Jay learned how to fix everything, all while working full-time and filling the roles of husband and father. We all learned to simplify, to use less power and water, and to deal with discomfort and mishap. Living aboard forced us to slow down, to appreciate small things, and to be more flexible. These were all skills we would need in the travels to come.

People have asked us a lot of questions over the years, like, why did we leave our comfortable life in suburban Atlanta? And once we decided to leave and embark on an adventure, why did we buy a boat only to keep it in a marina? After we had finally begun to travel and live a life of freedom and exploration, how could we be happy spending a season tied to a dock or sitting in a boatyard? Or, after we'd spent time in a place, made connections, and grown roots there, how could we decide suddenly to leave, braving wind and waves and an unpredictable future? These actions may look contradictory if you don't understand the yin and yang of boating life.

Zaid K. Dahhaj explains this principle of Taoist philosophy in his 2018 article published on Medium, "How to Practically Understand Order & Chaos." "The trick is within the balance of chaos and order. Too much novelty, everything degenerates. Too much order, everything petrifies. You want to be right on the proper edge…If you're completely in chaos then you're terrified. If you're completely in order then you're bored." The boat satisfies the need for both predictability and change. We can't rearrange the furniture inside, but we can change the view outside. While staying in one place for a while gives us the chance to dig deep, the novelty and uncertainty of travel keeps life interesting.

The boat offers a balance between joining a community and experiencing new things. I love to be a part of something, to join people in common pursuits—whether it's a vegetable co-op, sailing center, homeschool group, beach cleanup project, music jam session, or book club. But after several months in one place, doing the same thing every week, I begin to feel restless. Wanderlust is like an itch—it doesn't go away until you scratch it. Moving the boat gives me the opportunity to meet new people, participate in new ventures, and see new places, while at the same time offering the chance to return to a place and peel back another layer. We can stay as long as we like, go when we're ready, and plan a return visit.

We bought a previously owned, custom wooden catamaran not just because it was in our budget, but also because we wanted something unique, something we could work on, give new life to, and make ours. Jay especially likes the challenge of solving problems; the boat provides endless brainteasers and opportunities for improvement. It has been a labor of love, and though repairs can grow wearisome, I hope we never finish because making order out of the chaos brings us a lot of joy.

Our children have also had to find their own balance between order and chaos, as homeschool provides opportunities for both structured and creative learning. They have had the freedom to explore their own interests while also meeting the demands of daily schoolwork and chores. They are learning how to navigate relationships—meeting new or old friends, spending time alone or with a family member, relying on coincidental meetings or planning social engagements. In learning to organize their own small living spaces, they have had to choose carefully what they value, what they have space for, and what they can do without.

The reason the life afloat has been satisfying and sustainable for our family is that it requires us to walk the balance beam between work and play, travel and rest, new discoveries and familiar places, society and solitude, comfort and discomfort, creative mess and underlying order. Shipshape has come to mean more than just the state of the vessel; it is our own quest to order our lives while remaining open to the unpredictable.

# 11

## SEE WHICH WAY THE WIND IS BLOWING
### DECISION MAKING

*June 2015. It's three in the morning and we are sailing south past Miami. It's beautiful from a distance, like a neighborhood lit up for Christmas. I've just come up from below, where I've been sleeping during Jay's watch. He's awakened me because a small fishing boat has hailed the Coast Guard for assistance and we are the nearest vessel. Jay needs me to steer a wide circle around their bobbing boat while he communicates with the crew, ascertains whether they are in danger, and relays messages to the Coast Guard. Their battery has died, and they are anchored offshore in a heavy chop. They want us to give them a jump to recharge their battery—as if we were two cars in a parking lot instead of two small vessels bouncing around in the dark! The Coast Guard wants to know if the people are in distress or if the boat just needs to be towed in. Once we establish that they are not in imminent danger, that they still have working communication devices, and can call TowBoatUS or a friend to get back to port, we wish them luck, sign off with the Coast Guard, and head on our way down the coast toward the Florida Keys.*

*We have finally left Fort Pierce, where we've completed the season's projects on* Take Two. *We had hoped to sail to the Bahamas for a few weeks, but three of the kids' passports have expired, so we will have to get new ones and plan a trip for after hurricane season. As usual, plans are constantly shifting. We've been on the dock longer than we thought, missed our chance for spring sailing in the Bahamas, and now we're facing hurricane season. Should we travel north up the coast to get out of the hurricane box drawn by insurance companies? Should we go hide out on the west coast of Florida in a more protected bay or river? Or should we go hang out on mooring in Marathon in the off-season, touch base with our friends, catch some lobster, take a trip to the Dry Tortugas, and just run for it if a hurricane comes? I know that some people live and travel using maps and calendars—choosing a location, circling a departure date, preparing the boat, and leaving on schedule. But that's not us. Our plans bob around like that fishing boat in the dark, waiting for morning to shed some light on the situation.*

*Just a few days ago, sitting in our cockpit, we were untangling the complicated problem of where to go and when to leave. Ben was a witness to our endless deliberations and if-then statements; our back and forth made him chuckle. "How can such experienced, intelligent, capable people be so indecisive?" Though we didn't know the exact destination or departure date, we both had a clear sense that it was time to get off the dock, to stretch our sea legs again, and to chase new horizons. So now we are sailing down the coast overnight with no clear plan beyond spending a few months in the Keys with friends and lobstering this summer, hoping it is a stepping stone to a larger voyage.*

---

Many times we don't know if we're leaving until after we've left. We have learned from experience that sometimes the conditions we find on the ocean are not what we expected, and we've changed our minds and our direction so many times that we don't even call to say goodbye to friends or family until we're out in open water. In fact, if we tell people we're leaving, it pretty much guarantees that we're not. We basically wait for all the stars to align and then go for an experimental sail. Like a groundhog checking for sunshine in February, we creep out slowly, see which way the wind is blowing, and then decide.

We used to be planners. We had a house with a calendar on the wall and dates circled. Our week revolved around Jay's work schedule, homeschool co-op, church, and regular social engagements. We knew where we would be for holidays and birthdays. We took vacations and had travel itineraries. But we also had a wild hair. We wanted adventure and spontaneity. By definition, you cannot plan these things. You can have an idea, try it out, see how it goes, and decide if it works or if you need a new idea.

We had already bought the boat, so we knew the answer to what we wanted and why we wanted it, but the answers to where and when are always elusive, like a Fata Morgana mirage floating on the horizon, a thing of mystery. If you ask us where we will be for a season, the answer is, "it depends." If friends or family want to come visit us on the boat, we tell them they can pick the time or the place, but not both. We've become completely noncommittal, not because we're flaky, but because we really don't know. The more variables added to the equation, the less certain we become. All we know is that at some point, we will leave the place in which we find ourselves and sail to another place. If you ask, "When are you leaving?" I might answer, "On the right day, and not one moment sooner."

. . .

We love south Florida. We grew up in a town sandwiched between the Gulf of Mexico and the Everglades, the sea of grass, with its long-legged birds, alligator calls, and hammocks of cypress, pine, and mahogany. We like the smell of salt air, the sparkle of water, the feel of sand between our toes, and the peace of mangrove-lined bays. We both vacationed as children in the Florida Keys, which are close enough to the mainland that you can be in bustling Miami in a couple of hours, but far enough away to feel like you're on a tropical island retreat.

*Take Two* often stops in Marathon for a few weeks or months on the way to or from somewhere else. It's a small town in the middle Keys with things to do on the water—snorkeling, kayaking, sailing, fishing, and lobstering. Over the years, we've left and come home to a group of homeschool friends who live there, and we return often enough that the kids consider it their hometown. Though the boat is normally on a mooring, we have occasionally tied up in a marina, where dockage can get expensive. This is why we have often eyed waterfront property in the islands, thinking how nice it would be to have a place to keep *Take Two* when we're not traveling. We've also considered that we may want a place to settle down after the kids are grown. But property values, despite occasional wind damage and flooding due to passing hurricanes, continue to climb.

In the summer of 2015, after a date night at Herbie's Bar and Chowder House, we went for a drive in Marathon to look at a piece of property that Jay had spotted online, just out of curiosity. It had a canal, backed up to a state park with a wetland reserve (mostly mangroves), and was under foreclosure. It was undeveloped and overgrown, and the view, which might not be attractive to someone building a vacation home, looked wonderful to people hoping to find a hurricane hole for their boat. It was listed in a price range we could afford. We drove by again a week later, got out of the car, and walked around. And then the conversations started.

"What would we do with a piece of property?"

"We don't have to build on it right away. We could put in a dock for *Take Two* and run power and water. We could build a solar array. We could have a reservoir for rainwater, or a tank for fuel. We could build a little cottage."

"If we decide we don't want to build on it, we could use it as a 'nest egg.' With property values climbing, it could be a good investment."

"Or we could build on it. It could be a place for family to visit, or a place to help kids get started when they're ready to leave the boat, or a place for a parent to stay."

"Can we afford to buy property and travel in the Caribbean for the next few years? If we buy and clear property, will we run out of time with our kids to take a long cruise?"

And that's where the conversation sat. Even with unlimited funds (which we did not have) we would still run out of time. Time is a universal limitation. Rich or poor, we all have twenty-four hours in a day, and we never know how many years we have left to pursue our ambitions. Nothing shows us the speed at which time flies like raising children. One minute they are crawling around on the floor, scrounging around for choking hazards, and the next they are driving away in their own car. We have only a few short years to enjoy our kids between birth and independence, to invest in relationships that pay lifelong dividends.

This is why we left the suburbs of Atlanta and moved onto a boat: to raise our children at a slower pace, to homeschool, to travel as a family. We already felt successful. Having miraculously stuck with our five-year plan, we had traded our land life for a cruising catamaran. We had already crossed the Gulf Stream, made overnight passages, and spent time island hopping. The travel was the gravy to the meat and potatoes of living aboard. But we had bought the boat to do more—to go farther, to get out of our American bubble, to be more than tourists. Were we still committed to that plan? We stood at the crossroads asking ourselves that question.

We are not people who take commitments lightly. If anything, we approach life too seriously and overthink decisions. That might be why we're so slow to act. We consider every angle, run worst-case scenarios, and weigh risks and benefits. Why? Because once we do something, we're not likely to turn back. When we have heard stories of impulsive couples who sold their house, bought a boat, and sailed across an ocean in a single season, we are incredulous. Having a large family compounds our decision-making because we have to factor in the impact our choices will have on our kids; it makes us even more cautious adventurers.

If we originally bought *Take Two* to travel to the Caribbean with our family, and then decided to pause for a year or two to purchase property and build a dock, we would have to accept the risk of not breaking free again, and we would have to content ourselves with the voyages we'd already made. On the other hand, if we decided not to buy the property and stick to our plan of traveling to the Caribbean, we would have to accept the lost opportunity to purchase affordable waterfront property in a place we love. We were again faced with the decision to play it safe (use the time and money to build a nest egg) or take a risk (use the time and money for travel). Both looked like good choices for different reasons, and without the benefit of hindsight, we struggled to balance risks and rewards.

. . .

Making decisions is never easy. Humans are not omniscient. They possess the limitations of time, space, and chance, not to mention fallibility. We often want something but don't know why. Or what we want changes over time. We are motivated not only by our own passions, but by our fears, our pride, and our desire to please others. And we are limited by time, money, and circumstance. When two people engage in decision-making together, the variables—and differing viewpoints—increase.

Over the years, we have developed problem solving strategies that help us make big decisions. One is making an old-fashioned pros and cons list, but with a twist. We add a point value to each advantage or detractor, usually on a scale of one to five, with one being a not-that-important consideration and five being of-critical-importance. It helps prioritize a list of conflicting variables. Another strategy is risk-benefit analysis. My favorite version of this comes from Tim Ferriss, author of *The 4-Hour Workweek*, something he calls "fear-setting." Instead of goal-setting, he takes his worst fears into consideration and separates what he can control from what he can't, and whether he can mitigate risk. He weighs the consequences of not just trying and failing but the costs of doing nothing. We understand this in simplified terms, a mantra we developed in the days of boat shopping: "regret is worse than fear."

Sometimes we have to pause, take a deep breath, and review what we want out of life. We must prioritize, or reprioritize, goals and dreams. Do we still want the same things we wanted when we started out? Have we changed our minds about the short or long-term goals? What is best for each of us as individuals so we can be productive and successful? What is best for our relationship? What is best for our children? How do we balance conflicting goals? No wonder we feel exhausted after date night discussions and family meetings! But without asking the hard questions and making the tough choices, we might wander through life allowing incident and accident to determine our course. Instead, we pray, talk, debate, change our minds, pray again, and act only when we're certain we're heading in the right general direction. Even when we feel we've made a mistake, we debrief and recapitulate, make course corrections, and learn from our failures.

When deciding whether we should buy property in the Florida Keys, we reexamined the goals we had set when we bought *Take Two* and faced our fears. We had accomplished some, but not all, of the objectives. We wanted to expose our children to different cultures and languages and made that a high priority. Despite our uncertainty about the future, fear of rough weather and offshore passages, reluctance to leave our family and culture behind, we knew that if we didn't at least try to get to the Caribbean, we would always wonder *what if...?* We also recognized that we were running out of time. Our oldest son had turned fourteen in July of 2015. He would not be with us much longer; if we wanted to travel with the whole family, we would need to do it soon. In light of that fact, travel now ranked higher than security later.

Perhaps if we had bought that property in the Florida Keys, the story would end here. We would have put down roots, built a dock, a cottage, a solar array, a rainwater collector, maybe even a garden. We might have weathered the direct hit of hurricane Irma in 2017—we might have gambled and lost. Maybe we would have traveled, maybe even to the Caribbean, but with teenagers, it's more likely we would have stayed close to Florida long enough to get them independent instead of making a long open-ended voyage as a family of seven. As with all the choices we make in life, who can say what we could or would or should have done?

We came very close to ditching our original plan for an idea that seemed good at the time. Looking back, of course, I wouldn't trade any of our travel memories for the relative stability of owning a piece of land. Because we did not act on the opportunity to purchase the island property, we were free to keep moving. Sometimes inertia has affected our decisions; by not choosing, we were still making a choice.

Instead of calling the real estate agent, we ordered a new sail. The mast had fresh paint and the rigging was new; we now had a bowsprit and a crane at the top of the mast, so we made space for a Code Zero, a large, lightweight sail that would fill the gap between jib and spinnaker. We installed new solar panels on the recently completed hardtop. Our old dinghy, a twelve-foot AB fiberglass-bottomed inflatable, had begun to leak air and was unable to plane with the seven of us, even with the two-stroke Mercury twenty-five horsepower motor at full throttle. So, we purchased a new, thirteen-foot aluminum-bottomed dinghy and a thirty horsepower Suzuki four-stroke, the equivalent of buying a new family car. We also replaced the tired, overworked, Spectra watermaker with a much higher output double-membraned Cruise RO unit, which Jay installed himself. He also installed a new bank of lithium-ion batteries because our Lifeline AGM batteries, though still operational, were at the end of their expected lifespan. *Take Two* felt like a brand new boat, ready for indigo water and distant shores.

The crew seemed ready too. The older kids were able to take watches, help with boat chores, and feed themselves, if necessary. Our youngest crewmember was now four years old, no longer a baby. Jay and I had worked hard getting the boat upgraded and organized. We had found ways to make life and travel run more smoothly. Despite the difficulty of leaving a familiar place and people we loved, we knew we should finish what we had started. We had tested the waters, held a finger up to see which way the wind was blowing, and decided it was time to leave the safe harbor.

## 12

## PLUMBING THE DEPTHS
### GRATITUDE AND AWE

*April 2016. It is the last night of a weeklong passage from George Town, Bahamas to the east side of Puerto Rico. I am on watch in the wee hours and the sky is dotted with millions of points of light. The squalls of the previous days have disappeared and the seas have lain down until they are almost glassy. The mainsail has been zipped in its stack pack, and we are motoring gently across the Puerto Rico Trench, the deepest part of the Atlantic Ocean. I feel strangely alert and excited, anticipating our arrival in the Caribbean the next day. I walk the decks, checking on things and enjoying the beautiful night. It is not unlike the one so long ago that inspired me to try this lifestyle in the first place. The lights in the sky are mirrored on the surface of the water, but there is something else too.*

*Holding on to the lifelines, I look down into the water. Beyond the surface, I see a kind of green explosion. I have to adjust my focus to see through the glimmer of starlight on the surface, but soon, I detect a continuous eruption of these greenish lights, at all levels, going down and down into the deep. I realize the whole trench is full of this greenish light, which bursts and then fades like submarine fireworks. I knock on hatches and try to get the sleepy kids to join me* on deck to see this phenomenon, but I get few takers. "What? You woke me for bioluminescence!? Seen it a hundred times." Only Eli, the oldest, rouses himself to come have a look. He is not disappointed and is even ready with a scientific theory: bathypelagic sea creatures must be coming to the surface to feed. I think they might be jellyfish. Whatever the explanation, I have never seen anything like it. These are not the tiny green sparkles we leave in our wake but glowing balls of green fire.

It is easy to think of the ocean as the deep, dark, scary place about which we know so little, and which contains countless mysterious horrors. It is as foreign an environment as outer space, and just as inhospitable to us two-legged, two-lunged oxygen breathers. But it is not dark. In the abyss, where sunlight is but a distant memory, bioluminescence is the only light available. And, apparently, it is plentiful. I have read that three quarters of sea creatures can emit light, and that bioluminescence is generated by a certain kind of bacteria living symbiotically with other organisms and communicating using a chemical language to create bursts of light. This phenomenon may have a scientific explanation, but the neon glow still seems miraculous. I spend the rest of my watch staring down into the water, mesmerized. The next day, we wake to see a blue-gray cloud on the horizon, and it gets steadily bigger all day until we realize that the cloud at which we have been staring is the volcanic island of Puerto Rico, our first landfall in the Caribbean.

A plumb line, or lead line, is a device used since ancient times to take depth soundings and prevent a ship from running aground in shallow water. A lead weight is tied to a marked string and dropped in the water to measure the depth. In days of old, tallow or bees wax on the end of the lead would pick up material and give mariners clues about bottom type and location. Nowadays, boats have electronic gadgets that use sonar and digital displays to report soundings. We rely on nautical charts and a depth sounder to keep us off shoals and inside channels. When heading offshore, we can watch the numbers creep up until we drop off the edge of the continental shelf—somewhere around six hundred feet, when the display simply goes blank. In this way, we are literally plumbing the depths.

In the Bahamas, the Tongue of the Ocean is a deep submarine canyon that separates the islands of Andros from New Providence. To get to the Out Islands to the east, one must cross the Northwest Providence Channel. One moment, you're looking at a sandy bottom through twelve feet of aquamarine water and the next, the bottom drops away to six thousand feet and the water turns indigo. Once, while motoring across this trench on a windless June day, we decided to put the engines in neutral and go for a swim. The water was like the surface of a mirror—smooth, reflective, and crystal clear. We floated out a polypropylene drift line and an inflatable toy and jumped into the deep end. It was all fun and games until someone suggested putting on masks and snorkels. We put on our gear and looked down into the dizzying depths. It was literally breathtaking, more like flying or falling than floating. I felt like an insect on the surface of a pond, so small and vulnerable. Our kids thought it was great fun, but I had to get out and catch my breath. Sometimes plumbing the depths leaves me awestruck, experiencing equal parts fear and wonder.

I have never been so aware of my own insignificance as when sailing offshore in the middle of the night. Without interference from land-based electric light, the stars are bright enough to light the sky and the whole surface of the ocean. At the same time, I have never experienced such a feeling of connection with God—

my understanding of the intangible and spiritual source of all Life. This Creator is at once beyond, and greater than, the tiny blue marble on which I find myself an even tinier speck, and also a close companion that whispers peace and serenity to my unsettled soul. How can I explain this feeling of being at once infinitely small and also noticed? Embracing a spiritual experience requires one to become comfortable with mystery and paradox.

We were sailing south in the Atlantic toward Puerto Rico—the night before the one I have just described, in fact, and there were dark clouds mounting up all around, periods of rain and gusty winds, and the boat was rocking and waves slapping and shoving us around. Squalls are not unusual. We try not to sail if they are forecasted, but sometimes, like bullies following you home from school, they find you anyway. We had reefed the mainsail to slow the boat and try to make the motion more comfortable, but still, it felt a little out of control. One's perception at two in the morning after a few days with poor sleep may not reflect reality; we were, in fact, in no danger, but it felt dangerous and scary.

We were all asleep upstairs, the motion too uncomfortable for the forward bunks and the noise too much for the aft bunks. I was on watch, but I was growing increasingly anxious. When the weather is bad, watch schedules go out the window and we just make do, swapping whenever one of us gets weary. Jay came to the rescue and took over for a few hours while I crammed myself in a corner and tried to rest. I have never wanted off the ride so badly in my life—I realized that the only way to make the motion stop was to go for a swim. These are not sane thoughts, of course, but misery can give one tunnel vision.

And then I heard a voice. The voice, I should say, the one that called my name on a curving mountain road one winter when I was in my twenties and saved my life, the one that set my nervous mind at ease the morning of my wedding, the one that hinted as I stepped aboard *Katie Rose* that this kind of life would be my future—the very same one that I have heard at nearly every crossroads when I didn't know where to go or what to do. It is not my own inner voice—that one usually sounds like a broken record echoing back to me my own anxious thoughts—no, this voice speaks thoughts diametrically opposed to my own: counterintuitive, and yet, inexplicably, always right. This is neither the voice of reason, nor the voice of insanity. I have no scientific explanation for its existence, so I have concluded that it comes from outside the dimension I live in, even as it speaks somehow inside my own head. In the middle of troubled waters on a dark and stormy night, this voice said, clearly, "Look up." So I did.

My eyes had been squeezed shut, and I was huddled under a blanket trying to stay out of the wind. I looked up through parting clouds and saw a small clearing of sky: midnight blue with a sprinkling of white points of light. It wasn't much—the rest of the sky looked billowy gray and menacing—but it was enough. Somewhere, beyond my limited field of vision, the universe was going on like clockwork, the same way that it had been for millennia. Above the clouds and my temporary discomfort, all was as it should be. Through the stillness of these stars, I got the message loud and clear, and I felt an unfathomable peace. I sensed that everything would be okay, and I slept for a few hours until I was ready to take a watch so that Jay could rest. Without leaving my predictable suburban life to experience risk and danger and fear, I might never have known what, and who, was out there. And, at the same time, in here.

Arriving in Puerto Rico with our family was immediately rewarding—we knew we had stepped out of our comfortable cruising zone and into a new place. After an eight-day ocean passage, we arrived on the eastern side of this mountainous island, all green slopes and curves. It was in many ways what we expected—resorts, beaches, tourists. But in other ways, it was surprising. Puerto Rico gave us our first glimpses of volcanoes, waterfalls, and rainforests.

We rented a car and drove all over the place. We went to El Yunque National Park, hiking and swimming in La Mina waterfalls. We saw the caverns at Camuy. Crossing the Cordillera Central, we looked down into valleys filled with banana groves and family farms perched on steep hillsides. Descending the misty mountains onto the dry grasslands of the leeward side of the island, we witnessed the rain shadow effect firsthand. We spent time in Old San Juan, walking across the green meadows to the Fort of San Felipe, with its view of the wild Atlantic shore. We loved all of it. And it was only a taste of what was to come in the rest of the Caribbean. When we left Puerto Rico, our first stop was Vieques—the place where we swam in bioluminescent water. A mere day sail from Vieques, we arrived in the Virgin Islands.

Hiking, swimming, sailing, snorkeling—we did it all. The dramatic backdrop of turquoise water, rocky islands, and palm-lined beaches makes the Virgin Islands (both the U.S. and the British) a popular vacation spot. We were there in the off-season, thankfully, so we avoided crowds and could still find a quiet anchorage, which we often shared with our friends aboard *Abby Singer*. The enormous boulders of Virgin Gorda's Baths were a favorite with our crews. *What natural forces dropped these house-sized rocks here?* we wondered. We hiked together to the top of Salt Island, dove together on the numerous coral reefs, and kayaked together into partially submerged caves. In the deep bay of Peter Island, Eli and I snorkeled along the shore and discovered a whole ecosystem existing on a submerged rock. The longer we hovered, staring through our masks, the more intricate and tiny creatures we observed. *What makes all these plants and animals coexist in this microenvironment?* Everywhere we explored, there were beautiful, inexplicable, extraordinary things. We were filled with wonder when we really began to look at the world around us.

What to do with all this wonder and amazement? We can't seem to contain it—it makes us giddy and bubbles to the surface as joy and gratitude. *Thank you for this day—for the breeze that pushes against the sails…thank you for this starlit night…thank you for this crystal clear water and the colorful life that exists beneath the surface…thank you for this sunrise and another day on planet Earth.* I have heard even the most hardened atheists express these feelings, even though they have nowhere to direct their thankfulness. Gratitude is interesting because it requires an object—when we offer thanks, it is to someone. If someone gives me a gift, I say thank you for the gift. But when we feel this overwhelming sense of privilege to be alive and to be witnesses of the wonders of the universe, to whom do we direct our gratitude?

For me, it is a driving question, a lifelong quest. And so far, no other explanation that I have pondered besides *someone who loves beauty put me here* makes sense. Even in its postindustrial state, the world is too gorgeous for words. Living on a boat has put me near the pulse of this living universe, and I have learned to love not just the creation, but also the Creator, and to feel loved in return. Spending time in an old-growth forest, swimming in icy waterfalls, free diving with whale sharks, hiking up a volcano to see newborn rocks tumble down from a lava fountain: these moments have overwhelmed me and my family. We have felt full to bursting.

The natural overflow of this gratitude is twofold: creativity and conservation. Love is both something we express, and something we act on. As humans, we seem to be hardwired to create: we make music, sing, and dance; we tell stories and write poetry and plays; we draw, paint, sculpt, and take photographs; we decorate, design, and invent. To me, that's what it means to be created in the image of God; since God is Spirit, then it's in our spirits that we bear resemblance. Rachel Hollis echoes this sentiment in her book *Girl, Wash Your Face* when she writes, "creating is the greatest expression of reverence I can think of because I recognize that the desire to make something is a gift from God."

Similarly, a love for the creation inspires a desire to care for it. As homo sapiens—a species possessing self-awareness, reason, the power of speech, opposable thumbs, and possibly even a soul—we have the power, and arguably the responsibility, to shape not only our own destiny, but that of the ecosystems in which we live. We ought to be good stewards of nature's gifts, but most of us have proven to be very bad managers of our domain.

Here is the sad truth: humans have both the power to create and to destroy. That's what free will bestows—the power to choose to do what's right, to love, and to protect. It also implies the choice to do the opposite. And we certainly do choose poorly; as soon as we make a technological discovery—fire to cook our food for example—we use it to destroy, burning down the neighbors' village or the forest we live in and depend on. Part of our nature is to explore and discover, to tame the wilderness, to turn raw materials into something useful. As a species, we don't merely hunt and gather; we cultivate, we herd animals, we build villages, towns, cities, roads, civilizations. We synthesize and create new things out of old things. But in the process, we overconsume, we mismanage, we pollute, we fight over resources, and we lay waste.

And not until it's too late do we say, "Maybe we should set aside some land for the animals to live on, or some nature to enjoy. Maybe we shouldn't destroy every acre...every river...every beach...every forest." So we draw a boundary line around the wilderness, set up a ticket booth, and charge admission, taking advantage of the demand for natural beauty. People leave their neighborhoods, which might be urban or ugly or polluted, and flock to a nature preserve, national park, or wildlife refuge, where they spend a day or a week, trampling the wilderness, scaring away animals, dropping plastic water bottles and potato chip bags, and generally ruining the pristine beauty they came to see. On the other hand, if people don't spend time outdoors, they lose all connection with nature, and then they won't have the knowledge or desire to preserve it. What a conundrum!

What is the solution? Are we supposed to return to our agrarian roots, to the dependence on natural cycles, to small self-sustaining communities—to the myth of the noble savage? Are we to reject modern conveniences that use resources and create pollution, like electricity, transportation, and modern building materials—and go back to an idealized preindustrial era? Or is there a way to use our human ingenuity to learn to live more harmoniously with the planet—as symbiotes instead of parasites?

I have seen both mindsets in our travels: big resorts or cruise ships that create massive amounts of pollution and cater to tourists who want an escape without understanding their impact on the local environment, and a growing number of eco-resorts where visitors can immerse themselves in a habitat, eat locally-grown food, and experience beauty without destroying it in the process. The difference is in outlook and purpose. And, of course, resources—how can we make access to wilderness affordable and desirable for the masses? I want it to be possible for everyone to have a mystical, life-changing encounter with the natural world that would affect their decisions when they go back to civilization. It requires an attitude of humility and respect, and though that begins with education, it must include a measure of authentic connection with the natural world and maybe even a spiritual experience, which cannot be manufactured or sold.

Heading south through the Exuma islands of the Bahamas before the passage to Puerto Rico, we revisited some of our favorite cays in the chain. We are used to seeing flotsam and jetsam on the windward side of the uninhabited islands, but the amount of plastic trash has become a shocking sight. The rocks are covered—and I mean covered knee deep in places—in the detritus of our industrialization: bottle caps, flip flops, fishing nets, hard hats, broken toys, Styrofoam coolers, plastic bags, Coke cans, monofilament line, drinking straws, milk crates, forks and spoons, water bottles, and the colored confetti of plastic whatnots broken to bits in the sun but still indestructible.

On small islands all over the world, there is a constant influx of items packaged in plastic. And after consumption, there's nowhere to put the wrapper. Bernadette Bernon, former editor of *Cruising World Magazine* wrote about the problem in her log of Ithaka. She noticed that islanders were accustomed to throwing their waste in the ocean, but that in past generations it hadn't been a problem because all the waste had been organic. A banana? Eat it and throw the peel in the sea. The sea takes care of it. A coconut? Drink it and toss the husk in the sea. A chicken? Pluck it, cook it, throw the bones in the sea. But what do we do with this soda can? Or this potato chip bag? You get the idea.

And even on islands that have grown smarter, where they collect the trash rather than letting it go into the ocean, people burn large hills of organic and inorganic waste indiscriminately. My son Eli, a.k.a. Captain Vocabulary, dubbed it *garbecue*. Nothing like enjoying a sunset drink in the cockpit and having the gray haze of a trash fire drift by. This solution, of course, is no solution—the resulting microplastics enter the environment as air pollution and still make their way into the water and into the food chain, often as hormone-mimicking, gender-bending chemicals. There is no clean corner of the world because we humans are all in the same boat; we breathe the same air, and we drink the same water. And because of a combination of industrialization, globalization, and consumerism, we eat the same food, throw our trash in the communal heap, and ingest the same chemicals.

Assuming that you are not ignorant of the problem, or so poor that survival trumps conservation, reading that the Pacific garbage patch is at least the size of Texas—synonymous with BIG—might make you feel depressed. Crippling despair is one of the possible responses. It says, "This problem is too big to solve," which often leads to hopelessness and apathy, which in turn say, "This problem is beyond my control, so why bother?" These responses are, of course, a flush of the cosmic toilet, leading to the planet's downward spiral. And many have felt the pull of its swirl.

But there is another possible response: Hope. One year, after returning from the Bahamas, where our kids had witnessed the plastic problem for the first time, I was walking down the sidewalk with my daughter Sarah, who was five at the time. I am normally a conscientious litter-gatherer—I don't walk by trash; I pick it up and carry it to the nearest garbage can. But I had begun to feel jaded, like Holden Caulfield unable to erase all the bad words in the world, and I walked past a piece of rubbish. Sarah let go of my hand and said, "Wait, Mommy! Trash!" She turned around, bent over, and picked it up. As we walked toward the nearest garbage can, she asked, "Why didn't you pick up the trash?"

"What's the point?" I asked. "I can't pick up all the trash in the world."

"Yes, but you could pick up that piece!"

I was properly chastened. I had set an example that my young daughter had followed, and now I needed to take my own advice. One generation's actions have a ripple effect. Not only can we make the world better by behaving responsibly, but we can inspire another generation to care. This small gesture gave me hope. And Hope says, "We've conquered other big problems, and we can tackle this one together too." With Hope comes Belief and Action, which say, "I am a part of the problem, so I can also be a part of the solution." As an educated human, one born in a place where I have the luxury of focusing on conservation and not just survival, I am uniquely called to care, to communicate, to choose my responses, and to lead others toward responsible behavior.

My profound love of the ocean and hatred for the ubiquitous plastic I see there moves me to say that this is a good place to start. I am probably not going to take on the industry, fight city hall, or rewrite the laws. (Although, really, shouldn't we be holding big companies accountable for pollution and motivating them to switch to sustainable packaging?) But I am going to commit—as a consumer—to vote with my dollar, to set an example, and to share my convictions. I am attempting to rid our boat of single-use plastic—using glass storage dishes or mason jars for food, washable silicone bags instead of Ziplocs, mesh bags for produce, and heavy-duty fabric bags for carrying groceries. We carry water in washable stainless bottles or buy drinks in recyclable glass. We take our own washable plates and cutlery to potlucks and picnics. I cook as much as possible from scratch using raw ingredients instead of buying convenience items that are often packaged in plastic. These changes are possible on a large scale, but more people must be motivated to change.

My friend Julie, who started the American Bee Project to help save apis mellifera from colony collapse disorder, is a lawyer who argues that bees are an agricultural product and makes farmers eligible for tax breaks when they let land lie fallow as bee forage. She says that conservation must line up with economics. You can see this in Costa Rica, a country that has figured out how to monetize a clean environment by growing ecotourism, where it pays to preserve the natural world. While we were visiting the popular Eastern Caribbean island of Antigua, a law went into effect that made plastic supermarket bags illegal. Even in Guatemala, a country whose roadsides are littered with plastic, the grocery store where I shopped began selling grocery bags if you forgot to bring your own reusable ones—creating a financial incentive for conservation. This fills me with hope; while I used to be an oddity with my reusable bags at the store, almost everyone has them now. It may be only a drop in the bucket, but it's symbolic of an important shift in thinking.

Had I never chosen the simpler life—rejecting North American consumerism that demands that I buy more and better stuff, had I never seen the dreadful results of the exportation of our fastfood culture, had I never smelled burning plastic, had I never witnessed the floating islands of trash in the ocean, maybe I wouldn't care as much. But living on the boat has taught me something very important about natural beauty: we need it. We need the sun on our faces, clean water, fresh air, living food, moon phases, forests filled with birdsong, beaches, starlit dark skies. We need these things like we need good relationships with people. We need the wilderness because it is a sanctuary where we can interact not just with creation, but the Creator. We need it more than we need conveniently packaged snacks. And if we lose it, we may not be able to get it back, so we need a lot more people to recognize their need, too, and to care about saving it.

Reducing plastic waste—with the ultimate goal of using as much reusable and biodegradable stuff as possible, is a good starting place. In addition, simple living, self-sufficiency, and sustainable energy (smaller living spaces, solar panels, water catchment, wind generation, and box gardens, for example) are other possibilities which I had not considered before I lived on a boat but now see as good goals for everyone, not just sailors.

I'm so glad we went outside. Our family needed to discover wild and beautiful places together. Richard Louv says that "we cannot protect something we do not love, we cannot love what we do not know, and we cannot know what we do not see. And touch. And hear." The gift of living outside the house-sized box in suburbia is getting to know the Creator through the creation and learning to love it enough to preserve it. I recognize the constellations that mark the seasons, I have seen meteor showers in dark skies, I have witnessed the power and beauty of an electrical storm from front row seats, and I've smelled fertile soil and flowering plants miles out to sea when approaching land. I've come close enough to touch sea creatures, visiting birds, and strange insects. And, most importantly, I've plumbed the depths and climbed to the heights with my children, to whom the future of our planet belongs.

## 13

# SHIPS PASSING IN THE NIGHT
### FRIENDSHIPS AFLOAT

*June 2016. We drop anchor in Little Bay, the only place remaining for yachts to anchor near the island of Montserrat, part of the Lesser Antilles chain, often referred to as the "Emerald Isle of the Caribbean." Much of the city of Plymouth, on the southwestern coast, was destroyed by the Soufrière volcano in 1997. Two thirds of the island has been rendered uninhabitable, and the volcano is still active. We are here, in fact, to visit the Montserrat Volcano Observatory, to witness the power of molten rock firsthand. And we are meeting up with friends we haven't seen in several weeks, to share the experience with them.*

Abby Singer *drops anchor near us sometime in the night. We wake up to find them neighbors again, as they have been in several anchorages of the Bahamas and Virgin Islands. We met Andrew, Summer, and their two girls, Paige and Sky, in March, at the beginning of this journey. And while we aren't really buddy boating, we keep finding ourselves in the same anchorages as we make our way south to Grenada for the summer, and we are now intentionally trying to meet. When we see them again, it's like a family reunion. It's hard to believe that we've only known them for a few months, but sharing a cruising life means we've had a lot of time after school and chores every day to hang out, share meals, go on adventures, and build close relationships.*

*Jay and I find a friendly taxi driver named Moose, who takes us to the local Digicel store to get a SIM card so Jay can access the internet for work. We ask him about a tour of the Montserrat Volcano Observatory and the edge of the exclusion zone, where we heard you can get a peek at the damage inflicted by the volcano. He says that he and his friend Cecil could plan a trip with our two families for this afternoon. And he happens to know a great place where we can share dinner afterward (Moose's Place). This is a small island, and many people emigrated after the volcano erupted. Moose himself has moved twice—once after the first eruption and again after they redrew the exclusion zone borders. The people who stayed are hardy, friendly, and resourceful. They are, as Moose says, "pos-i-teeve!"*

*The day is perfect: we visit the observatory and learn everything we can about the eruption, even getting a view of its smoldering sides. Then Moose and Cecil tell us their stories as we stand on a lookout behind an abandoned, overgrown hotel outside the exclusion zone. We're standing in a rectangle which, upon further inspection, turns out to be a swimming pool, filled to the brim with ash and overgrown with weeds, overlooking the ghost town. A river of hardened pyroclastic flow runs through the center, and boulders the size of houses sit next to the ruined airport. Church steeples stick up out of a gray crust. It is disheartening, to say the least. Though only nineteen lives were lost, more than half of the population fled the island, and the capital city ceased to exist. Unlike the destruction of St. Pierre in Martinique a century before, when Mount Pelée erupted and killed thirty thousand souls, most of the people in Plymouth were saved because the volcano's warning signs were heeded. But lives were disrupted, families were displaced, and a way of life changed forever.*

*This does not stop Moose and Cecil from making us feel welcome on their island. Moose lives behind his bar and restaurant near the waterfront. He drops us off there and goes back out for fresh burger buns while Summer and I get busy in the kitchen, starting dinner preparations for our group, which has swelled to sixteen with the addition of Moose, Cecil, and another boat family on* Vidorra, *Bruce, Lauren, and their son, Luke, who we met in Nevis. We have a fun evening together in an atmosphere of easy camaraderie, as if we have all known each other for ages.*

It can be unsettling to sail at night. The body's system of balance relies partly on visual cues that disappear when the sun goes down, so sometimes we feel disoriented by conditions that seemed comfortable during the day, in addition to the sleepiness of missing bedtime or waking up in the dark for a watch. Not being able to see obstacles, shoals, markers, other boats, or shoreline features means relying on instruments and knowledge of navigation lights and the Convention on the International Regulations for Preventing Collisions at Sea (COLREGS). Sometimes other vessels, like the Disney cruise ships, are lit up like Christmas and easy to avoid. Others, like small fishing vessels on the Cayman banks, are practically invisible. Sailing at night requires extra vigilance and a hand-bearing compass, to make sure paths don't cross. It is not uncommon to hail a passing vessel on the VHF radio to make sure one boat is aware of the other. Boats that plan to make passages together may keep the radio chitchat up all night to keep each other company through the long, dark hours.

When we chose to live a transient lifestyle, we had to accept the loss of normal avenues of friendship. There is no gossip over coffee breaks at work, chatting over the backyard fence with the neighbor, committee or club meetings, or sporting events. It doesn't mean we are not sociable, only that gatherings are impromptu and serendipitous. Whoever happens to be in the anchorage will be invited to the birthday party, for example, and the crews of boats traveling the same route might congregate while they wait for a weather window to the next destination. Like ships in the night, sailors make friends in passing. Coincidence throws us together, and if we make a connection, we may have only a short time to spend enjoying each other's company, so we make the most of it. The result is what one might expect: lots and lots of acquaintances who we may never meet again, and intense close relationships that end up being lifelong keepers. The same is true for friendships ashore—everywhere we go, we find commonality and companionship among the members of our human family. Our paths cross, overlap for a while, and then go off in separate directions. It is a bittersweet way of life—always saying hello and goodbye in the same breath—an aloha kind of existence.

When we bought our first house in that Atlanta suburb, we dreamed of sitting on the front porch in the evening, going for walks, and visiting with our neighbors. We thought it would be like a 1950s sitcom. We were wrong. Sure, we went for walks, rode our bikes, sat on the porch, did yard work, and even made friends with some of the neighbors. But we noticed that everyone was so busy all the time and when people came home from work in the evening, they turned on the TV, and an eerie flickering blue light would emanate from all the windows on our street. We had chosen not to buy a television, so I guess we were the weird ones. When we bought our house in Florida, we were attracted to the neighborhood because it had a dog walking club and held block parties—it was a close community and had a homeowners' association. There were even some other stay-at-home moms on the block. But still there were the empty front porch chairs and the eerie blue glow.

It wasn't until we bought our boat and moved into a marina that we found the fellowship we had been looking for; people actually sat out on deck, or set up chairs on the dock, held potlucks, and met to play dominoes or cards together. It was the ideal community: sailors and powerboaters alike sharing a common love of the water, living side by side, helping each other when necessary, and holding weekend parties that lasted from Friday happy hour to a Sunday morning Bloody Mary. We were the only couple with kids, but they welcomed us into their group, kept an eye on our small people, and treated us like family. We aren't big partiers, but it was nice to be able to put the kids to bed and have some time with other adults when we were feeling sociable. Everyone on that dock had a nickname, and we were dubbed "the Robinsons." That was our introduction to the boating community, and to this day we keep in touch with some of the members of the H Dock Yacht Club.

The first year we sailed to the Bahamas, we spent six weeks in George Town, Exuma. There we met other boating families from the U.S. and other far-flung places, and the cruising kids would gather under the Casuarina trees near Chat 'N' Chill on Volleyball Beach, swinging like monkeys, digging in the sand, or building forts out of driftwood. There were bonfires and volleyball games, and kids coming over to *Take Two* to jump off the high dive and swim while grown-ups held cold drinks and chatted in the cockpit. It was wonderful. And then we said goodbye as families dispersed, some sailing north and others south and east to the Caribbean. But everywhere we went, this social scene would re-emerge, sailing families meeting up for a season, spending every afternoon together, swapping homeschool ideas, hosting movie nights and beach parties and bonfires.

Time is at once an enemy and a friend: not knowing how long the boats will be together in the same place means that people must move rapidly past the small talk stage, but having a similar pace of life means that there are hours each day to spend getting to know the people around you. Cruising creates dense friendships—more love packed into a smaller space and time. It's one of the things we enjoy most about our floating life: travelers become our friends and friends become our family.

Because the community of people who travel on their boats is relatively small, we often run into the same people in different places. Once, when we pulled into Royal Harbor near Spanish Wells, Bahamas, I noticed a small red monohull that looked familiar. I hopped in the dinghy and putted over to get a closer look: sure enough, it was *Lammeroo*, unmistakable. I tapped on the hull and called out, "Robin! Robin!" My friend Robin, whom I had met in the Florida Keys the previous season, popped her head out the companionway hatch like a prairie dog with a priceless look of bafflement on her face. She looked at me hard for a moment until recognition replaced confusion. This kind of thing happens to boaters all the time, and we say to ourselves, *I know that person...*or *that boat...*or *that voice, from somewhere... but where? And what name and boat goes with it?*

My favorite story of ships literally passing in the night happened when we were resting on anchor at South Water Caye, at the edge of Belize's barrier reef. Our friends April and Jacob and their two girls were on a passage between Mexico and Guatemala. We spotted them on the AIS and thought we might be within radio range, so we hailed them on channel sixteen on our VHF, the universal distress, safety, and calling channel.

"*Lark! Lark! Lark! Take Two.*"

"*Take Two*?! This is *Lark*! Switch to channel sixty-eight."

"*Lark, Take Two* switching to sixty-eight."

"*Take Two,* are you there? Over."

"*Take Two* here. It's a bit scratchy, but you're coming through. Where are you headed? Over."

"I can't believe you hailed us! We're on our way to Guatemala. How did you know we were here? Where are you? Over." My friend April's voice resounded with surprise and joy—they had no idea we were nearby and had not expected to be hailed by a familiar voice while on an overnight passage offshore.

"We're in Belize, waiting for weather to go out to the atolls. We were just thinking of you guys today and decided to look on your blog and see where you were. We searched for you on AIS and saw you would be passing close enough to hail. What are you up to? Over."

April's voice echoed scratchily over the radio. "We just bought another Amel after our last one was wrecked in Hurricane Irma in St. Martin. We haven't even changed the name to *Lark* yet, so we were surprised to hear you hailing us. We're on our way to the Rio Dulce for hurricane season. How about you guys? Are you heading south? Over."

"Sadly, no! We're heading north toward Mexico. Over."

"We just came from Isla Mujeres! I wish we could see you guys—we miss you! Over."

"We miss you too! Maybe next time." The voice grew faint and was replaced by static. I transmitted once more, hoping she would hear me. "Send me a message when you get there! Happy sailing! Over."

No response.

"*Take Two* back to one-six." They had sailed out of range, but not out of our thoughts. It had been as much fun hailing them as jumping out of a closet to startle a sibling.

Later, they returned the surprise. When our plans changed, we ended up in a marina in Rio Dulce, Guatemala. As we pulled into our slip, who should come running down the dock to meet us? The crew of *Lark*—Jacob, April, Audrey, and Lorelei, who had coincidentally come across the river for lunch at our marina's restaurant that day. We hadn't seen them in five years, since the year they had moved onto their first boat in Fort Pierce, Florida, where we had been neighbors at a marina. The friendship picked up right where it left off, and deepened and broadened, as friendships often do when given the time and environment to flourish. Our daughters, Rachel and Lorelei, who had been only three years old when we last met, recognized each other immediately and were soon like peas in a pod.

After leaving the Virgin Islands in May 2016, our friends on *Abby Singer* had gone to St. Martin to do some repairs, while we opted for the road less traveled: Anguilla, St. Eustatius, and Nevis. Anguilla had beautiful beaches and fantastic restaurants, but no kid boats. St. Eustatius (Statia) had a gorgeous national park with a great volcano hike, but no kid boats. Nevis, sister island to St. Kitts , had botanical gardens, hot springs, and interesting history field trips...and one other cruising boat. By some wonderful coincidence, the catamaran, *Katta 3* of Sweden, had three kids aboard, two teen daughters named Alice and Bianca, and a young son, Karl. Anders and Katarina had crossed the Atlantic with their family, cruised for a season, and decided to sell the boat in the islands. They were meeting a potential buyer for their boat in nearby St. Kitts .

We hit it off right away and were soon spending part of every day at the beach with them or sharing a meal on one or the other of our boats. Like our kids, Karl loved Lego, a common interest which made the language barrier negligible. We celebrated Swedish midsummer with them (as far as it is possible in the Caribbean), but they promised us a real celebration when we come visit them someday in the land of the midnight sun. It might happen—the world is small and we love to travel. But then again, it might not—our parting after a few short weeks together was tearful because we didn't know if we would ever see each other again. This is what it means to live like this: we are always sailing away from friends we love, towards friends we haven't met yet, coincidentally running into old friends again, and then parting once more. We have adopted the farewell phrase, "see you next time," because we never know when our paths may cross again.

When we met the crew of *Abby Singer* again in Montserrat, we were so happy to see them that we decided to sail on together —to round the island's south side for a view of the exclusion zone from the water and a better sailing angle to Antigua. After our tour of the island and dinner at Moose's Place, we left early the next morning in a brisk breeze, the wind filling our sails and pushing us along at a good clip. As a large catamaran on a beam reach, we were faster than *Abby Singer* and soon way ahead. Between the islands of the Eastern Caribbean, the prevailing wind and ocean currents find an unencumbered entrance to the Caribbean Sea—a lot of air and water come funneling between the mountainous islands to create exciting conditions for sailors. We euphemistically call these island hopping passages sporty or salty, but we have headed into the gap between two islands and found wind and waves so unpleasant that we have changed our minds, retreated to a protected anchorage, and waited for a better day.

As we rounded the southern tip of Montserrat, the first ocean swells met us head-on. It wasn't bad enough to merit turning around, but I did feel a sense of responsibility to our buddy boat. I called *Abby Singer* on the VHF and warned them to stow any loose items because it was going to get rough. Aboard their smaller monohull, they were going to feel the motion more than we did. And we would reach the anchorage and be settled in hours before they'd drop the hook behind us in the dark.

This is the downside to buddy boating, the practice of traveling in a group. There is a fine line between safety in numbers and herd mentality. We've witnessed groups of cruisers meeting to discuss departure plans where one boat is responsible for weather, another for route planning, and still another for reading the cruising guide about the destination. This always makes my lone-wolf husband uncomfortable. He values self-sufficiency so much that he often won't even share his opinion with another sailor about the weather—the single most popular topic of conversation among captains. He doesn't like to take or give advice about route planning, storms, or forecasted wind speed and direction. Though he doesn't mind making friends while sailing, he would rather just meet coincidentally in the same harbor than plan to depart or arrive in conjunction.

Two or more boats traveling together may give the illusion of safety, but in an emergency, there is often little one buddy boat can do for another. Convoys of boats are used to prevent piracy, but if one boat gets boarded by armed strangers, the other boats can do little more than report the attack to the Coast Guard. The boats feel responsible for each other but bear very little actual responsibility; the crew of each boat must deal with circumstances on their own. It is nice to have company out there, to have someone to chat with on the VHF about conditions or to alleviate the boredom, but having that familiar sail on the horizon may not be worth the sacrifice of self-reliance. Boats that travel together make decisions based on what the group wants, not necessarily what's best for each boat or crew. Decisions should be based on good judgment, not social pressure, because conditions that are tolerable for one boat may place another boat in danger.

As captains of two very different boats, both Jay and Andrew always did their own weather planning, and the crews agreed that we would always make decisions about passages independently. As a result, we rarely traveled on the same day and sometimes arrived at different destinations. We might take the same weather window to move south, but after the run from Montserrat to Antigua, we never jointly planned passages. We would spend a few fun-filled weeks together, and then go our separate ways before meeting up again elsewhere. In this way, we leap-frogged down the Leeward and Windward Islands of the Lesser Antilles until we landed in Grenada for hurricane season.

The boating life can also be lonely. There are seasons when we don't meet up with any other kid boats, or when boats are crossing paths, heading opposite directions, and the time together is abbreviated. We rely largely on coincidence to meet friends, and sometimes we're in different places at different times. Our kids have grown up literally all over the map, moving every three to six months, and not every personality is suited to making friends quickly or to saying goodbyes repeatedly. The opportunity to homeschool while traveling and to see different places and walks of life has a downside: isolation and transience prevent a person from developing normal social ties. This is not always a bad thing. We have avoided some of the pitfalls of a typical American upbringing. Our kids are not socialized, but they are sociable—able to have conversations with adults and heterogeneous groups of kids. They are also their own social group: with fewer outside friends, they have had to be each other's companions.

But I also recognize the emotional impact of uprooting our kids and dragging them all over God's green earth expecting them to make and leave friends like changing their clothes. This has especially impacted Sarah, who is more introverted, and who doesn't like leaving her small group of close friends in the Keys. Broad horizons are great, but sometimes, it's comforting to return to a place we call home for a taste of the familiar.

One of the benefits of living full time on the boat is that we can spend weeks or months in one place if we like or return to a place we passed through to spend more time. For example, Marathon. Every time we find ourselves in Boot Key Harbor, we touch base with our land friends and the relationships grow stronger. In this way, though we are transient, we have put down roots and formed lifelong friendships—a rare and precious treasure.

Another, though more challenging, way to develop long-term friendships is to meet where cruisers gather for hurricane season. During the winter and spring, boats spread out, finding their own pace and exploring their own corners of the world. But seasonal weather forces them to gather in safe havens where they are less likely to be found by tropical cyclones. On the East Coast of the United States, boats head north for the summer to the Chesapeake, Maine, or Canada, for example. In the Caribbean, they gather in Luperón in the Dominican Republic, or the islands of Grenada, Trinidad, and Tobago, or farther west in Panama and Rio Dulce, Guatemala. The likelihood of running into old friends and cruising kids goes up when the boats congregate in these so-called hurricane holes. The potlucks, open mic nights, dominoes games, and movie nights are like fairy gatherings in a forest; they reappear and disappear as seasons change.

In November of 2016, we made our way north to St. Lucia and then took a downwind ride on the trade winds to Bonaire, Caribbean Netherlands. We were on our way to the Western Caribbean, and we were hoping the crew of *Abby Singer* would come with us. After a wonderful season in Grenada, where we were neighbors on the dock at Le Phare Bleu marina, and where our kids had cemented their friendships while sailing, swimming, playing games, hiking, and jumping in waterfalls, we knew we had found friends that would last a lifetime—kindred spirits who added the missing ingredient to our cruising life. Every experience was more memorable when shared.

They had gone back to the U.S. for a visit in October and were bringing us back a part for our broken oven, but we had to wait for them to fly back to Grenada, prep their boat for travel, and find good weather to sail west. We held Thanksgiving off for a week, wanting to wait for them and share it together in a beautiful place. We anticipated their arrival, coached them on the VHF as they entered the bay at midnight, even went over by dinghy to help them pick up their mooring and give them a squeeze. And we did have an amazing month in Bonaire with them and with the crews of *Penny Lane* and *Delphinus*. Holidays spent far from home can be lonely affairs, but we became home for each other, sharing old holiday traditions and making some new ones too.

At the end of December, we took a three-day weather window for a sail to Santa Marta, Colombia. This is a tricky passage as the wind and seas in the area of the Guajira Peninsula can be intense. We saw fifteen knots of boat speed and fifty-knot gusts as we sailed downwind, surfing down ten-foot waves. And that was in good weather. We were worried about *Abby Singer*—how would they handle the passage? They needed a week of good weather to make the same trip, and it was close to the time when the

Christmas Winds make westward travel difficult, if not impossible. We had said, "See you next time," and were hoping that even if they didn't make it to Colombia in December, maybe we'd see them in Panama in January.

And then came the email. We all cried when we read it. The coming of the New Year had brought bad news: *Abby Singer* was sailing back to Florida. A financial setback had dried up the cruising kitty, and they would need to get back to sort things out. They had taken a year off from their busy lives—a year that transformed their family and ours, too, and the memories would last forever. But when they weren't able to join us, we felt like someone had died. We couldn't imagine sailing on without them. And that's when we learned that opening ourselves up for friendship also means leaving ourselves vulnerable to heartbreak. But the opposite is true, too: if we shield ourselves from pain and close our hearts to our fellow sojourners, we also shut out the opportunities for love, joy, and fellowship. Meeting friends, no matter how temporarily, is like passing a ship in the night on the vast ocean—we are comforted to know that we are not alone.

# 14

## TROUBLED WATERS
PATIENCE

*December 2016.* It is the first day of our passage from Bonaire to Colombia. We've studied the route and the weather and decided that it's now or never. We felt more trepidation than normal about this passage because we have heard it can be challenging and uncomfortable, not to mention that there's no turning back. But today the wind is light and from behind us, and the sea state is comfortable. It's been a while since we've used Don Johnson, our asymmetrical spinnaker, so we raise the sock and watch as the peach, seafoam green, and light gray chute fills with wind and our velocity increases dramatically. At this rate, we can shut the engines down. Have we finally found the perfect spinnaker day?

Clearly, we have forgotten all the trouble this sail has given us. It is original to the boat and was badly rigged when we bought it, and more than once Jay has pulled out his knife and threatened to cut it loose. I usually send him back to the cockpit and patiently untangle the lines. Even after taking it to a field to lay it out, measure it, and rerig it, we still feel like it needs babysitting. All goes swimmingly at first, but then disaster strikes as we approach Curaçao. Flukey winds cause the spinnaker to fold in on itself, but before we can adjust the *sail angle or sheets, a gust from a different direction wraps the pastel colored nylon fabric around our forestay and furled genoa. We should have seen it coming. We swear like sailors.*

*We start by trying manually to shake out the twist, but the wrap is tight, and just out of reach. To make matters worse, the upper half of the sail is still full of wind, a balloon pushing us inexorably onward. We can't stop the boat because we can't snuff the sail. Wrapped as it is, we can't even cut it loose. If our arms were only twenty feet longer, we could reach up, give it a little tug and loosen the twist and fix everything. As we head into the lee of Curaçao, the wind goes very light. Jay's troubleshooting hasn't resulted in a successful strategy, so it's time for me to make one of my idiotic suggestions, so crazy that they sometimes work. Inspired by a dumb joke about turning the house to screw in a lightbulb, I ask, "Since we can't turn the wind back around the boat, what if we turn the boat around the wind?" If we can get one twist out, we might be able to use the sheets to untangle the rest.*

*So, we jibe the boat. We make a full 360-degree turn, and the balloon in the sail gets bigger, but that's because it's beginning to unwrap itself. With help from our able-bodied teenage crew, we begin the arduous task of switching the sheets manually from port to starboard and from starboard to port, until the sail is untwisted, and we can get the snuffer sock back down over it. With a sigh of relief, we head west, away from the Dutch Caribbean. We stuff the spinnaker back into a locker—muttering curses and letting Don Johnson know that he will not see the light of day again for a very long time.*

*A few days later, we round Cabo de Vela on the Guajira Peninsula of Colombia. The wind is now blowing a steady thirty to forty knots, and we are on a fast sleigh ride down ten- to twelve-foot waves with a double-reefed mainsail and no jib. We are seeing double digits of boat speed, which feels fast and tense. In the pink morning light, we catch a glimpse of the snow-capped peaks of the Sierra Nevadas, and by the afternoon, we are heading into the marina in Santa Marta. We call on the radio and they direct us to a slip. But the moron who designed the marina placed the docks perpendicular to the prevailing winds, making docking difficult, especially for a catamaran. Furthermore, the finger piers are not reinforced with pilings, so a bump on the end might break it away from the dock, something we find out firsthand. Despite line handlers on the dock, fenders on the boat, and bursts of engine power, Take Two's broad side catches the wind as we turn into the slip and we crunch against the finger pier, wrenching it sideways. Not the smoothest landing we've ever made, but we're here.*

*We are giddy with success. "South America…South America!" We say it over and over again, like an incantation. Despite all the trouble, we have somehow managed to arrive at another continent.*

No one said this would be easy. In fact, if we had wanted easy, we would have stayed in our house and read sailing magazines. Sailing is thrilling because it entails taking real risks. It's uncomfortable, wet, fast, dangerous, windy, and fun. As hard as a rough day at sea can be, the sense of accomplishment we feel after conquering—or surviving—the elements, creates euphoria and a willingness to repeat the experience. When we're out on the ocean, we must rely on our wits, our knowledge, our preparation, and our patience to see us through troubled waters. And when those fail, determination and the stubborn will to survive kick in.

But beyond the actual wind, waves, and weather, there are other difficulties to surmount. A lot can go wrong on a boat: things break, leak, or burn. Moving parts and instability can cause all sorts of injuries. Small mistakes can lead to sizable consequences. The ocean is unpredictable. Boats are unpredictable. Sailors are unpredictable. These are the reasons why we say that "all plans are written in wet sand at low tide." We never know what will happen. People who like this spontaneity go on adventures; those who do not, go on vacations.

Ironically, a successful life on a sailboat is not dependent solely on one's ability to sail. I've met people who couldn't sail but moved aboard and liked living on a boat. And I know people who are excellent sailors but who don't like the idea of living full time on a cruising sailboat. It takes a special skill set to live on a boat—flexibility, patience, a sense of self-sufficiency, the ability to forgive, and a penchant for fixing things. Yes, there are sunset drinks, but they usually come after a day of unclogging a head, troubleshooting a dinghy motor that won't start, or chasing down a stubborn leak. For those of us less mechanically inclined, the sunset drink is earned by cleaning the bilge, handwashing the laundry, or making something amazing for dinner—from scratch, with ingredients scrounged from under the floorboards. Living with the ocean's mercurial moods keeps us on our toes. We have to be ready to deal with anything that comes up, which requires resourcefulness and adaptability. And often there's no one to count on besides ourselves and our crew—and whatever supplies and skills we bring along.

When we got married, I didn't know that Jay had the capacity to become Mr. Fixit. And he didn't know that I could become a Galley Goddess. Neither of us had any formal training in sailing, boat repair, or first aid. But when we bought our first boat, we began our schooling. Jay learned what to do and not to do by making repairs and mistakes on *Blue Bear*. I read sailing books and went to a women's sailing seminar. Jay went to seminars and read books on heavy weather tactics. I taught myself to cook and bake from scratch. And after we bought *Take Two* and moved aboard with small children, we felt we needed to be ready for anything. We took a two-day class on first aid afloat—including how to do injections and stitches. I attended a Safety at Sea seminar and a weekend Ladies Let's Go Fishing event. When he could, Jay did all the renovations and repairs by himself, and when he couldn't, he watched the one doing the work so he could learn. I took a celestial navigation class, just in case our GPS ever fails (but God help us if it does).

And, of course, we went to the school of the sea, which is the best, though the most ruthless, kind of instruction. By trial and error, by hook or by crook, through hell and high water, we figured things out. And though we have a hard-won confidence, hardship can still humble us, and we have a lot left to learn. This learning is a painful—and memorable—process, requiring Herculean patience with ourselves and others. But if we can find an inner calm despite frustrating circumstances, it is like pouring oil on troubled waters, a trick the ancient mariners used to calm rough seas.

Boats break: it is a fact of life. Old boats, new boats, wooden boats, steel boats, cement boats, fiberglass boats. Things go wrong because they can; Murphy said so. And the corollary to his famous law predicts that they will do this at the worst possible moment. Sometimes it's an engine failure in a rocky channel. Sometimes it's an electronic chartplotter blackout as we approach a shallow reef. Sometimes it's a leak—and we have to ask the question, salty or fresh? Are we sinking or did we overfill the water tank? Sometimes it's an electrical problem—wires can melt, short out, or even catch fire. Always it feels like an emergency, but often a stopgap measure can remedy the immediate situation until we can address an underlying problem more thoroughly. Breakage reveals a weakness in the system—something a previous owner installed a long time ago that needs redoing, something that needs an upgrade, or something we didn't even know we should worry about. Anything short of a catastrophe makes us feel grateful—every time there's a problem and we solve it, the boat and the crew get stronger.

While we've received lots of good advice from other sailors and read many helpful books and articles, calling for help in the middle of a situation is rarely an option. If a boat near shore is unable to get to an anchorage or marina, there may be a towboat service. While we might feel embarrassed to have to call for this kind of assistance (assuming it were available), we would not call the Coast Guard unless someone were dying or we were stepping up to the life raft from the deck of our sinking ship. Ideally, we want to solve problems ourselves. They say that luck favors the prepared, and on a boat, training, routine maintenance, tools, and spare parts can save one's life and floating home.

Once, an engine shut down as we entered Fort Pierce channel in the dark. It is a gnarly and narrow entrance to the Intracoastal Waterway on the east coast of Florida, lined by two rock jetties. It's almost impossible to plan an entrance for slack tide, so there's usually a strong current running in or out. With one engine, our catamaran was having a hard time maintaining forward motion and steerage—the current was pushing us sideways. I was wrestling with the steering wheel while Jay went looking for the problem. He found it: our fuel gauges are notoriously inaccurate, and one tank had run low and the engine had slurped up the dregs and shut down. Jay had previously installed a fuel transfer pump so he could move (and filter) diesel from one side of the boat to the other. This was not the first time his forethought and preparation saved us, and it wouldn't be the last. He got the engine back up and running and we made it to the anchorage without further mishap. A few weeks later, a boat came limping in with a gaping hole in her side—a less happy run-in with the rocky channel.

Another time, when Jay was working on an ancient through-hull in the port head, a corroded sea cock broke off in his hand, and water started geysering in. Had this happened to me, I would have been in a complete panic. Thankfully, he has a cool head, and remembered the bag of wooden bungs of assorted shapes and sizes we keep in an emergency locker. He calmly grabbed the right one and shoved it in the hole to plug the leak and let the pump take care of the accumulated water. Once, when Jay was out of town on business, I tried to sink the boat—using a freshwater hose at the dock. I negligently overfilled a water tank and had to use emergency pumps to get the water out. I never want to see the floorboards floating again, but now I know what to do. (I later saved a friend's boat from sinking at the dock by checking the hose in the water fill port!)

Aside from the occasional wire meltdown, we've only had one fire scare. Our original engines were problematic and have since been replaced. They were twin twenty-nine-horsepower Volvos, and the one on the port side didn't fire on all cylinders, always smoking until it warmed up. While it didn't bother us too much, it caused a panic at the marina where we first lived aboard. It was Friday afternoon and we were preparing for a weekend at anchor. We had groceries, fuel, and full water tanks. We ran through our preflight checklist: warm up the engines, haul the dinghy, test the transmission and steering, get lines ready for easy cast off. By the time we pulled away from the dock, the sun was beginning to set. There was a commotion on the dock next to ours. Fire trucks had pulled up with red lights flashing and firefighters in full gear were running down the dock as we passed. We looked around for the telltale billowing black smoke—a boat on fire is dramatic, unmistakable, and frightening—but saw nothing. As we motored slowly past the restaurant on the pier, a few guys came out of the bar waving their arms and shouting, "YOUR BOAT IS ON FIRE! GO BACK TO THE DOCK! THERE ARE FLAMES SHOOTING OUT OF YOUR EXHAUST!"

Jay and I looked at each other quizzically. We have diesel engines with twelve feet of wet exhaust, so we were less likely to have an engine fire than we were to be attacked by a fire-breathing dragon, but just to be on the safe side, Jay checked the engines while I went down the back steps and looked under the bridge deck: nothing out of the ordinary. Then we connected the dots: smoking port engine, reflection of the sunset on the water, happy hour at the marina restaurant, firefighters. No way our boat was on fire, but maybe it looked like an emergency to someone who'd had a few drinks. *Take Two* continued motoring out into the Manatee River toward the anchorage. We shrugged as we passed the restaurant and waved goodbye to the happy-hour crowd. An hour later, a Coast Guard helicopter passed over and we wondered: are they still looking for a boat on fire?

We were sailing back from Anegada toward the North Sound of Virgin Gorda in the British Virgin Islands. It was a blustery day and we were racing along at nine-to-ten knots on a beam reach. Jay used to crew on a raceboat, so he likes a brisk day sail in nice conditions. It's even more fun if there are other boats with which to compare your vessel and skills. On this particular day, the boat was rocking and rolling, and music was playing—most of us were actually enjoying it. (It helped to know that a nice calm anchorage and good meal were awaiting us at the end of the day.) Our friends on *Abby Singer* were somewhere behind us, and we had planned to meet up to do a tour of the island the next day. And then Jay stepped out into the cockpit, just as the boat gave a lurch, and caught his littlest toe on the base of the captain's chair. He says he knew instantly that it was broken.

There was nothing he could do about it, short of popping Advil, putting his feet up, and icing the injury. But something as small as a misstep and a painful pinky toe can be completely debilitating. I had to take over all sorts of jobs that Jay normally did because putting weight on the foot caused excruciating pain. He couldn't even step in and out of the dinghy. He missed all the fun the next day because he couldn't leave the boat—the hiking up Gorda Peak, driving the North Sound Road, trying the local food (no spicy goat water soup for you!), touring the old copper mine, and swimming in Spring Bay and climbing around on the giant boulders. It took six weeks to heal and was a reminder of how pain can humble us.

This was nothing compared to the injury he did to his hands on the way to Bonaire a few months later. We were sailing west from St. Lucia with the big code zero sail in a downwind reach. When I was on watch one night, the wind picked up and I woke Jay because I thought we were overpowered. He agreed, and we decided to change sails. As he cranked in on the continuous line to furl the sail, I eased the sheet. This usually works when the wind is light, and the sail rolls up tightly and evenly. In strong wind, and without the main to blanket the flow of air, the belly of the sail ballooned as the ends wrapped tightly. As fast as Jay hauled on the line, the wind pulled out the wraps from the middle of the sail. In our rush to get the sail down, he had not thought to put on sailing gloves, but after we finally got the sail in, we looked at his shredded palms and I got out the first aid supplies. The next day, we caught two mahi-mahi while trolling in calmer weather. After Sam reeled them in, we realized that he would need help skinning and cleaning them. While other members of the family catch and clean fish, I usually take charge of cooking them. Not this time. I pulled out the fishing book and studied the diagrams while Jay supervised the process with bandaged hands.

With the kids, we've been very fortunate and have rarely visited a doctor's office or emergency room. We eat well, and get plenty of fresh air and sunshine, so we don't get sick very often, though seasickness is a common ailment. Aaron suffers the worst, and after trying every natural remedy known to man, we eventually decided to drug him every time we untie the lines. Because we participate in adventurous activities that invite injury, there have been a lot of bumps and bruises, most of which we've been able to treat at home. For the occasional serious accident, we've been close to shore and had access to clinics, thankfully. When Aaron broke his arm, we were in the Florida Keys and I drove him to the hospital. Sarah sprained an ankle once, and Sam chipped a tooth while skateboarding, but there was a dentist nearby. A couple years ago, Rachel had a close encounter with spider monkeys that ended in a visit to a clinic in Portobelo, Panama, where she got antibiotics, a tetanus shot, and stitches for a deep scratch.

And Eli? Eli swallowed a pop-top from a soda can the day before we were leaving for the Bahamas one year. We were having hamburgers and fries under the tiki at Burdine's in Marathon, and I had gotten up to take Rachel to the bathroom.

When I came back to the table, Eli informed me, with a straight face, "Mom…I just swallowed the top of my soda can."

Ever the practical joker, I didn't believe a word of it. "Yeah, right."

Eli fixed me with his stare, articulating each word and speaking slowly. "No, really. I swallowed the top to my soda can!"

I looked at the rest of the family to see if they were in on the joke. Jay looked slightly concerned. "Is he for real? You guys are terrible. How does one *accidentally* swallow a piece of metal?"

"No, Mom, he really did," the other kids contested. Until the rest of the family vehemently corroborated his story, I was ready to laugh it off.

I turned to him, goggle-eyed. "How on earth did you do *that*?" I asked.

"I was fidgeting with the top, and it fell into the can. I forgot all about it, and then when I tipped the can up to drink the last gulp, I felt it go down!" He looked like he was about to cry, whether from the startling incident itself or from my disbelieving him.

As the gravity of the situation dawned on me, I went into Emergency Mom mode. "Did it have sharp edges?" I turned to Jay. "Do we need to worry about internal bleeding? Should we take him to Miami Children's Hospital? Is he going to need abdominal surgery? I guess we're not leaving tomorrow!" My mind was suddenly a-buzz with worst-case scenarios. *Mayday! Mayday! We're going down!*

Jay could see me spiraling out of control. "Hey, Tanya—before you start planning the funeral, let's talk about the possibilities."

I sat down and took a deep breath. We took another top off a soda can and examined it carefully. It had mostly rounded edges. Maybe it wasn't as bad as I thought. But we agreed that we should certainly not go offshore without knowing whether it was going to pass smoothly through Eli's digestive system. We started strategizing a wait-and-see approach. We would need to find out how fast his body digested food and watch for warning signs. Perhaps he could pass the soda-top on his own, negating a long drive to the mainland and a traumatic and expensive hospital experience. We should know within a day or two whether it was serious or not. We decided we needed some kind of marker—something non-toxic, indigestible, and colorful that would help us determine how fast his system worked. And we needed him to collect his waste so we could find the small piece of aluminum. With five kids, I have changed a lot of diapers and seen some interesting things. *What do kids swallow that they don't digest?* I thought. *Corn?...Raisins?...Peas?...Crayons?!* I had hit upon a foolproof idea.

When we got back to the boat, we had Eli swallow some pieces of white crayon we cut up. We sent the next morning's results through a sieve. Then we cut up a blue crayon. By the following day, after swallowing pieces of a yellow crayon, he found the pop-top in the bucket he was using as a toilet. He came into the main cabin, grinning with relief and triumph. There had never been any sign of internal bleeding, and once we had the result we were looking for, we sterilized our bucket and sailed away, crisis averted.

Perhaps because we like to be prepared, perhaps because Someone Upstairs is looking out for us, perhaps because we have a great boat that takes care of us, we have had very few actual emergencies, though more near misses than I can recount here. Things do break, leak, and burn. Injuries occur. We can prepare ahead of time to be ready when things go wrong, but we can't always account for how people are going to respond in an emergency. Some people panic, some shut down, and some remain calm and levelheaded. We have done fire drills with our kids. We have practiced man-overboard procedures. We have docked and anchored more times than I can count. We check things before we leave for an offshore trip. We carry spare parts, more tools than we know what to do with, first aid supplies (including crayons!), and food sufficient for an army. We make our own water and power. We have a catamaran, with built-in redundancy.

But when something goes wrong, we just hope that one of us will have the wherewithal to remember what needs to be done. Survival at sea requires staying calm in a tense situation, accepting reality, readjusting to the new circumstances, and patiently problem solving. We all appreciate and look to our captain's cool-as-a-cucumber responses in the middle of a crisis. And over the years, experience has helped me to replace anxiety with confidence, and I have learned to take a deep breath and to pray instead of panicking. Being able to access mental peace in the middle of troubled waters, whether they are the literal or figurative kind, calms the nerves and helps us think clearly. Most importantly, our ability to surmount difficulties and exhibit patience sets an example for our children, who will face their own obstacles in life.

## 15

# ON THE RIGHT TACK
#### GIVE AND TAKE

*July 2017. We are volunteering in a one-room schoolhouse in an indigenous village in Panama. Though this hut bears no resemblance to the brightly lit, colorful, and air-conditioned public school classroom in which I used to teach kindergarten in an immigrant neighborhood of Atlanta, some things are familiar: the smiling brown faces, the glittering dark eyes, and the murmur of Spanish while I write words in English on the board. Our friends Bobby and Shirlene, expats who operate a small marina on Isla Bastimentos in Bocas del Toro, Panama, have gone back to the United States for a family visit, and in their absence, we're helping with the English program they started a few years ago.*

*Eli and Aaron have been working in Bobby's boat repair shop in the afternoons alongside the Ngäbe men, but this English class is my first interaction with the community beyond buying hand-extracted coconut oil from a mother and daughter I met on Red Frog beach. Sometimes one of my kids comes with me, but today I take the dinghy alone to the marina to walk with my friend Leah, who also lives on a boat, to the tiny elementary school in the village. We wear rubber boots, rain jackets, and long skirts*—boots for the walk through the muddy field which serves as a cow and chicken pasture, jackets for the rain which comes and goes without warning, and long skirts because we are more welcome if we dress modestly, since the natives were supposedly civilized by conservative missionaries.

The trail takes us along the base of the hill, at the edge of the rainforest. Morning mist hangs in the trees and the air smells of rotting leaves, rain, and smoke from cooking fires. On my shoulder, I carry a bag full of games, flash cards, and colored markers for playing Pictionary with vocabulary words. I also have my ukulele so we can practice Old MacDonald using popsicle stick animal puppets. We cross the bridge over the creek that runs from the jungle to the mangrove-lined bay, stepping over the space left by the missing board which serves as a makeshift bathroom for the children. Despite efforts by missionaries and nonprofit organizations to improve water quality and sanitation, the villagers are resistant to change.

The school is a two-room structure on low stilts with raw boards for walls and open spaces near the top which serve as windows for light and ventilation. As we approach, several smiling faces peek out of the doorway, and we hear excited murmurs. Inside, a group of five- to twelve-year-olds squirm in their seats and issue greetings to Leah and me in Spanish and English. I love being here, but it is also hard. In addition to experiencing culture shock, I wonder if what we are doing is really helping the Ngäbe people. Being able to speak English will give these children opportunities in the future to work outside the village, but the tourism industry which provides those jobs is also changing the land and destroying a way of life. These larger philosophical questions are pushed aside as Leah and I engage with the children, practicing greetings, answering questions, and beginning the day's lesson.

When sailing to windward, a boat tacks back and forth across the breeze in a zigzagging fashion. For a catamaran, which doesn't sail well upwind, it can mean going very fast in the wrong direction. But sometimes, when the wind is from the right angle to the course we have chosen and the waves are not too rough, we raise the main and unfurl the genoa and off we go like a hot knife through butter. Nothing feels better than a boat coming alive in the wind and leaping forward over the sea with a whoosh. For us, a beam reach is fast and direct, and if our course lies perpendicular to the wind, we are definitely on the right tack. This doesn't happen very often, of course, and the perfect conditions don't seem to last very long, which requires us to change tack and steer a new course.

The physical act of tacking has taken on a metaphorical equivalent: just like the wind, life's circumstances can shift, requiring a person to change tack. When we find ourselves heading in the right direction, our approach may be said to be the right tack. What became apparent the longer we traveled as a family, was that our experiences were changing us. Though we started sailing to have fun and see the world, our focus had shifted. Each place we visited and each person we met pushed us in a new direction, prompting us to stretch and grow. And we began to see a larger purpose in our voyage.

Besides making memories and enjoying natural beauty, another reason we went sailing with our family was to help our children learn to appreciate the privileges they were born with—a family that loved them; the opportunity to get a good education; plenty of food, water, fresh air, and sunshine; financial stability and a comfortable home. We never wanted them to take these things for granted, but to see that they were rare and special, a gift meant to be shared. We knew this instinctively, but travel taught us practically and experientially. First, we removed some creature comforts, which made us more grateful for luxuries like climate control, unlimited warm water, fresh food, and plentiful indoor space. Second, as we began to see how other people live, the lens through which we saw ourselves and our culture changed. Finally, taking it a step further, making friends cross-culturally showed us how little we understand concepts like wealth and poverty, how helping sometimes makes things worse, and how we have more commonalities than differences with the other human beings on this planet. We found ourselves on the right tack, building authentic relationships wherever we went, experiencing empathy, and feeling moved by compassion to do something for other people when we saw a need.

We have seen poverty in the United States. There's no big city or small town without the ubiquitous homeless person panhandling and holding a WILL WORK FOR FOOD sign at an intersection. We're not big believers in giving handouts, but we do give time, energy, and money to nonprofits and ministries that help poor communities. I used to volunteer at the Atlanta Union Mission, helping with childcare while mothers were in job training workshops, twelve-step programs, and Bible studies at a halfway house for homeless women and children. I understand that poverty is hard to define, that it is a complex problem, and that efforts to care for the poor are often backward and do more harm than good. But I also believe in sharing goodness; those who have ought to be generous with those who have not. The word *charity* carries a lot of connotations, but the idea is based on altruism and compassion: showing unselfish love toward one's fellow men. For me, it's the acknowledgment that what separates me from the one begging in the street is largely an accident of birth: I was born to a white, middle class, educated couple in North America in the 1970s and as a result, I enjoy privileges I can't truly appreciate because I take them for granted.

In the Eastern Caribbean, we were used to seeing the less fortunate, at least by North American standards: unemployed black men sitting idly outside small multicolored shacks with skinny dogs in the yard and laundry flapping in the breeze. Islanders accustomed to rich tourists were always sticking an empty hand in our faces asking for something. But even then, we were beginning to rethink our definition of poverty and our reactions to it. What does it mean to be poor? Do we have an obligation to help the poor? And if so, how? The answers to these questions have been debated over and over during our travels. Throwing money at a problem is like taking a painkiller for a chronic headache—it might alleviate the symptoms for a while but won't address the root cause of the illness. As an outsider, I may not even correctly perceive the problem, and my well-meaning efforts to help might actually make things worse.

But what we saw along the road to Cartagena, Colombia in January 2017 bore no resemblance to the poverty I had seen before. It left me speechless and overwhelmed. We took a bus trip from Santa Marta to the old walled city of Cartagena. It was a fantastic step back in time to the Spanish colonial era—fortifications, churches, history museums, and cobblestone streets lined with restaurants and shops. But on the way there, we had a different kind of cultural experience. We saw people living in absolute squalor in small roadside villages. Think Unicef poster poverty, dwellings literally made of trash: cardboard boxes, sticks, fabric, plywood, plastic tarps, sheets of rusty metal—anything strong enough to make four walls and keep the rain off. The surrounding yards were paved with garbage. Small, naked, and filthy children ran around with skeletal dogs. As we went over a bridge, we suddenly saw dozens of people running, biking, or riding motorcycles, donkeys, and tuk-tuks toward an overturned truck and trailer: it had lost its payload of ice and the local citizenry were losing no time in saving this luxury item before it melted away in the equatorial heat. I was getting my first glimpse at how the rest of the world lives, and it threw my own life into stark relief.

Socioeconomic disparity is often a roadblock to friendship, and we have tried to figure out how to get around it. I listened by the hour to the sob stories of the boat boys in the Windward Islands—they come in all manner of floating vessels to offer help to cruisers or to sell produce or fish or handicrafts—and sometimes just to get a little sympathy. Buy a pound of limes and get a lesson about the Rastafari religion for free! It takes time and patience, but careful listening and a willingness to make an exchange—recognizing that each has a need and each has something to offer—helps break down cultural and economic barriers. We made friends with a local taxi driver, Winston, in Grenada, when we hired him to give us a tour of the island when we first arrived. After that first day, we called him whenever we needed a ride, but he quickly became the go-to source for information about all sorts of other things—history, culture, holidays, and tropical produce. When he wasn't driving the taxi, he was working on his small farm. Our friendship and cultural exchange started when he brought me some noni-juice. In return, I made him some cashew apple jam. He then brought me some lemongrass for tea. It was a relationship born of a common love for food and drink—something which can bring people together and break down the walls built by culture, language, privilege, and poverty. Rich or poor, black or white, male or female, everyone has to eat.

One of the reasons we had left the safe harbor of our old life was to learn a new language and to meet people who weren't like us. We had certainly felt what it was like to be a minority while traveling in the islands, but as we slowed down and dug deeper in South and Central America, we began to peel back the veneer of tourism and discover what it meant to make friends with the locals. Though we had a lot to offer, we also had a lot to learn.

When we first arrived in Colombia at the end of 2016, I contacted a local family who were friends of friends, Silvana and Leo and their daughter Maria-Alejandra. They were more than generous with their time—Rachel and Maria Alejandra became friends, and Leo and Silvana were eager to show us the treasures of their country, starting with food, language, and natural beauty, and ending with friendship and generosity. Silvana began teaching me Spanish, and to get more time for practice, I offered to come help her pack boxes as she and Leo prepared to move to Medallín. They took our family hiking in Parque Tayrona on New Year's Day 2017 and helped me book a field trip to La Mina, where the kids and I, along with a group of sailing friends from the marina, toured a coffee farm and cacao grove, hiked through a bamboo forest, and jumped in a waterfall. I was blown away by the warmth and welcome they showed us—foreigners who were barely more than strangers.

I also made the acquaintance of the produce man, Gustavo, at the local supermarket where I shopped. I was practicing my kindergarten-level Spanish and learning about South American fruits and vegetables at the same time.

"*Hola, Gustavo! Cómo esta usted?*" (Hello, Gustavo, how are you?)

"*Muy, bien, Señora, y usted?*" (Very well, and you?) He responded, attempting to speak slowly for me.

"*Bien, gracias. Por favor, puede decirme que es esto?*" (I am well, thank you. Please, can you tell me what is this?) I asked, holding up fruit that looked like plum tomatoes on a branch. I had bought some the previous week, discovering that the only thing they had in common with the tomatoes that grow on vines was the shape and color. They were indigestible. I thought I was missing some vital information, hence my question.

"Ah! *Son tomates de arbol!*" his face broke out in a broad smile.

I translated in my head. *Tree tomatoes?* I asked another question in Spanish, attempting to mirror his accent, "*Y cómo se prepara eso?*" (How does one prepare this?)

He launched into a long description in beautiful Colombian Spanish, only a few words of which I understood, namely, *jugo* and *azúcar*—juice and sugar. No wonder we had not enjoyed them!

"*Gracias, Gustavo! Hasta la próxima vez!*" (Thank you, Gustavo! See you next time!) I thanked him and put the *tomates de arbol* in my basket so I could take them home and make them into a sweetened juice.

Each week was a repeat of this conversation. "*Que es esto?*" I would ask, and he would introduce me to some new and delightful South American fruit: *mangostino, lulo, granadilla,* and *uchuvas*. He also recommended we try some traditional Colombian foods eaten at Christmastime, like *natilla*, a custard-like dessert that Sarah and I attempted to make by following the directions on the box, written only in Spanish. He answered a myriad of other questions while I, in turn, helped satisfy his curiosity about the United States and how we had traveled to his country—all in my baby-Spanish, sprinkled with embarrassment and laughter. I was learning to speak without fear of making mistakes in a relatively forgiving environment. The kids had been practicing Spanish using Duolingo, but the real-world application of the language only began to dawn on them when I took each one into town on shopping field trips.

These early language lessons came in handy when we sailed a month later for Panama, where I became friends with Steici, who has a produce stand in Bocas Town. Her husband took the ferry and drove weekly to the mountains to bring back produce to sell on Isla Colon, Bocas del Toro. I also met her two children, and I would bring Rachel with me on shopping day so we could spend some time together doing a language exchange. This was mutually beneficial, as she needed help communicating with her gringo customers and I needed to improve my basic Spanish. Meanwhile, Rachel had a friend with whom to play and practice new words.

We were plugged in at a dock at Red Frog Resort and Marina for hurricane season 2017, resting and recuperating from a year of nearly constant travel. Jay was working full time and we were running the A/C to ward off bugs and heat. On the weekends we were exploring the many picturesque anchorages of the Bocas del Toro archipelago. We had been living there for several weeks, walking almost every day to the beach on the Caribbean side of Isla Bastimentos to swim in the afternoons, when we met David. We had seen him raking the beach and serving Coco Locos to resort guests but had never spoken with him. One day I discovered, using my newfound language skills to ask some very basic questions, that he was from Venezuela, that he was, in fact, a refugee, working in Panama and sending money to his family. In his home country, where everything was in shambles, he had been a medical doctor, but he couldn't get a license to practice in Panama, so he was raking a beach and taking emergency calls when resort guests needed medical attention. He hadn't seen his family in over a year. He had a wife and two boys, about Sam's and Rachel's ages—and he missed them desperately. I couldn't even imagine what that would be like, to make that kind of sacrifice to keep my family fed and cared for.

After a few conversations, I invited him to join our family for homemade pizza one Friday night, which is a family tradition. We had to cross a few barriers in befriending David. First, he worked at the resort, and I was a resort guest, so just getting him in the gate at the marina raised eyebrows. Second, he spoke no English, and my Spanish was, shall we say, *limitado*. Third, we come from drastically different cultures, something we discovered one mistake, and one laugh, at a time.

Remember what I said about food connecting people? It turns out that Venezuelans eat too! He began to come for pizza regularly and to play card games and dominoes with our kids after the meal. We quickly learned the word "*tramposo!*" (cheater!) and we all learned to throw it around. David not only liked to eat, but to cook as well. He began to hold informal cooking classes; he taught Sam how to make a sweet dessert rice with fresh coconuts, Arroz con Coco. He spent a day teaching Eli, along with the rest of the family, how to make Venezuelan tamales wrapped in banana leaves, *hallacas*.

At the beach, we commenced a daily English-Spanish language practice. We owe our comfort in Spanish largely to the time spent with David, a debt I could not repay with all the pizza in the world. As is so often the case, when we want to help another person—to reach out in compassion or friendship—it is we who receive the help.

Until we came to South and Central America, I'd had very little actual interaction with native peoples. Growing up in the United States, Jay and I had seen the Seminole Casinos in

South Florida and the Miccosukee Indian Village and Gift Shop in the Everglades, but beyond the ubiquitous Seminole-built tiki huts, we thought very little about the displaced peoples of America. It was during our time in Santa Marta and in the islands of Bocas del Toro and San Blas that we first understood what it might have been like if our forebears hadn't rounded up the natives and relocated them to reservations. Despite the Spanish conquistadors' efforts to enslave or eradicate native peoples, they failed to subdue many of the tribes they encountered in the New World.

In Colombia, we had friends who went on a guided multiday hike to the Ciudad Perdida, or Lost City, to see the ancient ruins of an indigenous settlement in the jungle. We investigated the hike but felt uncomfortable with the stories we heard—the interactions with native people along the track involved passing out lollipops to children and renting hammocks at rest stops. The Lost City is on Kogi land, and the Kogi are private people, like the jaguars for which they are named, protectors of the environment who do not support the sullying of their sacred places by tourists. The Tayona National Park closes once a year just so they can come down from the mountain and perform a ceremonial cleansing. Hiking to the Lost City seemed more like a trip to the zoo than an authentic experience—I didn't want to see the natives in their traditional garb, or to destroy their traditional diet by sharing our teeth-rotting sweets with their children. So we kept our distance. Once, I saw a Kogi man on the streets of Santa Marta, looking starkly out of place in his white robe and cap, and I wondered what he thought of the noise and pollution.

But in Panama, one doesn't have to hike up a mountain to see the natives. The Ngäbe people inhabit a huge area in western Panama, the Comarca Ngäbe-Buglé, one of Panama's five semiautonomous regions that have substantial indigenous populations which retain, among other things, land rights. This includes the islands of Bocas del Toro, where they ply the waters in dugout canoes called *cayucos*. They would knock on our boat in a quiet anchorage and ask if we had anything to trade or donate—like dry goods or children's clothing. Sometimes they were selling fish. When we rode the bus to Boquete to enjoy some time in the mountains, we passed through their lands—numerous small farms where they harvested bananas and kept cattle. We saw women in vivid dresses walking with children along the roadsides. A traditional handicraft, hand-woven string bags made from the fibers of a plant that grows in the jungle, were for sale in every craft market. Though I know the clash of cultures and civilizations always costs something, I was amazed to see the degree to which they were keeping language, culture, and a traditional way of life intact despite Panama's development and influx of travelers and expats.

Imagine multimillion dollar mega yachts docked a few hundred yards from a village where people are cooking over open fires and using the toilet in the mangroves. The disparity is mind boggling. But alleviating prosperity guilt by giving things to the Ngäbe Indians creates a cycle of dependence that does neither the giver nor receiver any good. At the same time, without intervention—education, sanitation, access to medical care, and job training—the villagers struggle to survive in a rapidly changing world. Which is why we found ourselves teaching English in the Ngäbe school. English is a key that opens a door of opportunity, even as it closes a door on the past. The children's grandparents typically speak only Ngäbere, but their parents also learned Spanish, and government operated schools in the villages are conducted in Spanish with English taught as a second language. If the language and traditions are not preserved and transmitted by elders, cultural identity will be lost, at least in the islands where tourists and expats have moved en masse.

While living in Panama, my friend Shirlene tried to organize classes for a cottage industry in handmade crafts. She had something to offer the natives—education and medical care—and the natives had something to offer in return: knowledge of traditional handcrafts, like the making of string bags. Some of the women in the islands have lost this knowledge, so Shirlene found a woman from the mainland who was willing to teach local villagers how to make the bags so they could retain a traditional craft and earn an income at the same time. This is the kind of help I can understand, something that benefits all the stakeholders. Meanwhile, Bobby was employing Ngäbe men in his boat repair shop, providing jobs for them and much needed labor for him.

A slightly different situation presents itself in the San Blas islands on the Caribbean side of Eastern Panama. This comarca is governed by the Congreso of the Guna Yala (Guna or Kuna people). It is popular with cruising boats because of its beautiful coconut-palm-clad islands, coral reefs, clear water, and peaceful anchorages. The natives are friendly, and it's possible to go far enough from civilization to feel that one has stepped back to a simpler time. We spent a month there at the end of 2017, after we sailed west from Bocas and made stops in Colon and Portobelo. We had met some Guna women at Shelter Bay Marina who came to display and sell their handsewn *mola*s. *Mola*s are colorful, multilayer, reverse-applique quilt squares using geometrical patterns or depicting local animals like toucans and parrots. They are worn by native women as panels on the front and back of a beautiful blouse and are sold to tourists from around the world as art for framing or for decorating pillows or handbags. I asked the ladies we met if they would be willing to show Sarah how to make a *mola* (a simple one, anyway) and, in exchange, I offered to use my sewing machine and fabric scraps to turn some of their *mola* panels into pillow covers which they could sell easily. It was an in-kind trade, skill for skill and time for time.

In San Blas, we met a woman who paddled out to our boat in a little dugout *ulu* to sell her *molas*. She introduced herself as Morales, and we spoke in Spanish, but her real name and language are Guna. She climbed nimbly into my cockpit with her five-gallon bucket of wares—beaded anklets, headbands, *molas*, and embroidered shirts—and I offered her a cold *limonada*. The Guna are small but strong; she was not much taller than my daughter Rachel but paddled her wooden *ulu* alone, upwind in choppy seas. Her husband is a fisherman—free diving and spearfishing for a living. She sells hand-sewn crafts to support her family—two grandchildren who were orphaned when their parents died (*"Mama en el cielo, Papa en el cielo"*). Not unlike the boat boys of the Eastern Caribbean, she came out to the boat nearly every day, trying to sell me the same stuff.

With no bank or ATM in the San Blas, I had to buy groceries with the cash we were carrying, so I usually began our conversation with, "I can't buy anything today, but you're welcome to come in and have a cool drink." She never begged, exactly, but she was sizing us up—we had things she needed, and she had things to sell. She started bartering *molas* and beads for household goods like towels and extra fins our kids had grown out of. Even after I had headbands for all six of my nieces and more *molas* than I knew what to do with, she kept coming by. Her grandson needed to go to the city to see a doctor, and she had to sell what she had to pay the fees. We made a final purchase and then sent her away. A cynic would say she took advantage of my compassionate nature to sell as much as she did. But while the line between *generous* and *sucker* is very thin, I prefer to give someone the benefit of the doubt than to be suspicious and stingy. If someone asking for help is dishonest, they will have to live with that, but I couldn't live with myself if I looked the other way when someone was in real need.

One of the most important things we have learned from life on the boat is that the one who offers help might soon be the one asking for it. In material and spiritual ways, we reap what we sow. As a Christian family, we try to live according to the Golden Rule, to treat others the way we would want to be treated; it is how we raise our children. In essence, this is what it means be on the right tack relationally, and to understand that it is better to give than receive—that by giving, we actually receive a greater blessing!

Giving isn't synonymous with reaching into our pockets—often it requires something with higher personal costs: a willingness to build relationships and to share time, energy, and skills, or to partner with those in the community who are already making a difference. Furthermore, I am a strong believer that "to whom much is given, much will be required." (Jesus, Gospel of Luke.) My gratitude for all we have overflows as hospitality, and encourages us to share it with others, whether that's inviting strangers in for a meal, giving things away, being on the losing end of a trade, or spending time helping someone.

If I had to choose the one way that travel has changed me the most, it would be that I now feel like a part of the larger human family. I see people who were invisible before: I befriend people who bear no visible resemblance to me, I imagine the place of origin when I hear a new voice, and I feel at home in the world outside my hometown. This family, like any other, offers love and a sense of belonging, but also has some dysfunction and requires self-sacrifice and hard work. I don't know if I would have learned what it means to be part of that family if I had stayed in the house with the white picket fence. And I know that our whole family benefited from changing tack and seeing the world through new eyes.

## 16

# SMOOTH SAILING
### SIMPLE APPRECIATION

*February 2018. We motor slowly past Morgan's Head, a boulder with a distinctive profile named for the infamous pirate who once haunted these parts. We look back at the green hills of Providencia and Catalina Island, waving and taking pictures. It is morning, the fresh breeze salty and clean smelling. It's time to move on, but we've had such a lovely experience that the goodbye is hard. The boats* Soul Rebel, Seahorse, *and* Aqua Lobo *are nestled peacefully in the anchorage overshadowed by the Peak of Grand Old Providence, but not for long. They are all headed south to Panama while we are pointing our bows northwards. We went out for a farewell pizza dinner last night—a group of eight adults and eight children, once mere strangers, but now, after three weeks in this cozy harbor, good friends.*

*I am also thinking of Luis and Tachi, islanders we have befriended—and how I may never see them again, though I hope we'll keep in touch. More than any other place we have visited on our voyages, Providencia embodies the idea of island paradise we had when we bought our boat. Imagine a small green volcanic island rising out of the turquoise sea, surrounded by coral reefs, rimmed by sandy beaches and palm trees. The people are friendly, but the place is so hard to get to by water and the airstrip so small, that it has never been mobbed by tourists, so it has retained its Old Caribbean charm. Its inhabitants are proud of this, and protective, but still welcoming; houses are painted in cheery colors, and even the bus stops look like candy colored cottages. There are little open-air seafood restaurants along the southern beaches, walking paths on the islet of Santa Catalina, a hiking trail to the top of the Peak, snorkeling on the reefs around Crab Cay, and golf cart rentals which offer an inexpensive way to get around—literally around, as it takes only forty-five minutes to circumnavigate the island. It is the most perfect place we can imagine.*

*The day I met Luis and Tachi, I had been sitting on deck, trying to capture the picturesque shoreline on paper with acrylic paint. They kayaked by and admired the boat. They had been out fishing, and I invited them in for refreshments. While we sipped tropical smoothies from my blender, I asked a few questions. Before long, we discovered a common love of music. Luis plays guitar and bass in a five-piece band that showcases local folk music, using instruments as diverse as a mandolin, a guitar, a horse jawbone, maracas, and a bass made from a wash tub turned upside down. His band is preparing for a competition in Cartagena, Colombia. I handed Luis the acoustic guitar we have on the boat and he worked magic on the strings. I played along (as best as can be expected from a middle-aged white woman who recently taught herself the ukulele) and we had a good laugh. The kinship was instantaneous. We made plans to meet again.*

*The next day, I went ashore to look for them in the place on the shoreline they had indicated. This is unusual. Locals don't normally invite us into their lives like this. I felt awkward as a crowd of brown-skinned children stared at me coasting onto the beach in my dinghy. I asked where I could find Tachi or Luis and they all pointed simultaneously toward the gymnasium, where basketball players were trickling out after practice. I saw Tachi and waved. We walked back to her house and I gave her the gift I had brought—the painting of the island I finished the day we met.*

*The next day, Luis offered to take Jay and the boys spearfishing. He said no visit to Providencia would be complete without a beach barbecue. Everything they caught in the morning would be cleaned and prepped for cooking over the fire that night. He told us to invite all the sailing friends and he would invite his friends and family and we would all have a good time. And it was everything they promised —an evening of firelight, good food, and friendly chatter. I snapped a picture of Little Luis playing my ukulele as the sun went down; he's a carbon copy of his dad. This is why we left a comfortable home on land and crossed large expanses of water: to make friends and memories. Without a doubt, Providencia will keep a little piece of our hearts.*

---

Sometimes nothing breaks. Nothing leaks. No one gets sick. Sometimes, our life is exactly like the slick pages of sailing magazines: umbrella drinks, sunsets, hammocks, and crystalline water. Sometimes we have perfect sailing weather and never have to turn on an engine. Sometimes we catch a big fish on a calm passage. Sometimes dozens of dolphins play in our bow waves, and we all stand on deck feeling how lucky we are to live on the ocean. These are not the times we tell stories about, but there are enough of them to balance out the other times, when everything seems to go wrong. Smooth sailing motivates us to keep going, to live through the unpleasant in order to catch a glimpse of the perfect.

I've skipped whole chapters about our trips to the Bahamas because they were perfect, beautiful, and unsullied by breakage or bad weather. What's to tell? We would leave Florida, cross the Gulf Stream, and sail across the Bahama Banks, where we island hopped down to the Tropic of Cancer and spent carefree days swimming with our children, catching fish, making beach bonfires. Once, we had the luck of sailing past Nassau with a brisk wind on our beam, but the waves were blocked by the island of New Providence—we glided over the water like a speed skater on ice. Usually a fast sail comes with uncomfortable chop and swell, but on rare occasions, smooth sailing is more than just a metaphor.

Another time, we anchored off a little island in the Exumas with no name written next to it on the chart. We dinghied over, played on the beach, circumnavigated the island, and swam in its turquoise waters. In the late afternoon, we gathered dry sticks and branches into a pile on the beach, went back to the boat to put together a quick pot of chili and pack S'mores supplies, and returned at sunset to build a fire. We had dinner and dessert and then Jay and I sat in beach chairs drinking wine out of plastic cups while our children danced around the fire like natives after a successful hunt. When the fire died down, we doused the embers and buried the remnants of our evening, packed stuff and people back into the dinghy, and putted back to the boat, leaving bioluminescent sparkles in our wake. It was our island. We claimed it for the private country of *Take Two*, penciling this name on our chart: Robinson Island.

One summer, we did a bottom job in Spanish Wells. It was trouble free, completed in record time while we rented a beautiful house and had a visit from Jay's mom. While there, we made new friends, took the kids cliff jumping, and celebrated Bahamian Independence Day with festivities at the community park including homemade coconut ice cream. We then crossed the notorious Devil's Backbone to Harbor Island in Eleuthera without a pilot to guide us between the shallows and the treacherous reef, and it was easy. We dropped the hook near Man Cay and discovered a beautiful, isolated coral head within swimming distance. I took three-year-old Rachel for her first snorkel and we saw a baby sea turtle!

I have dozens of memories like these, but they are not the stories that first come to mind. More often, we recount more exciting tales, like The Day We Battled the Cold Front, The Time We Dragged Anchor in the Night, The Afternoon We Encountered Water Spouts, or When Mom Got Stranded in the Dinghy. Like J.R.R. Tolkien writes in *The Hobbit*, "it is a strange thing, but things that are good to have and days that are good to spend are soon told about, and not much to listen to; while things that are uncomfortable, palpitating, and even gruesome, may make a good tale, and take a deal of telling anyway." Smooth sailing is all it's cracked up to be, but not much to talk about.

For three years, while we traveled south down the chain of Eastern Caribbean islands, then west across the top of South America, and finally north up the coast of Central America, our solar panels made usable electricity from the sun; our batteries stored the sunlight and turned it into refrigeration, lights, blended drinks, and hours of computer use; our generator ran on hot evenings and provided the luxury of air conditioning for a few hours; our watermaker gave us freshwater for showers, clean clothes, dishes, and ice; our washing machine desalted our clothes and made them smell nice; and our dinghy hauled all the groceries, took us to beautiful reefs, and even towed kids on wakeboards. After the years of deprivation when we first moved aboard, we felt like the Clampetts in Beverly Hills.

Except for Rachel, we all remember the early days on the

boat, the one-gallon-bucket-and-cup showers, six people sharing one hand pump, saltwater-flush toilet, lying awake on sweltering nights drowning in pools of our own sweat, washing clothes by hand in a five-gallon pail and wringing the water out, throwing away produce that we couldn't keep cold enough, and moments of frustration as the family car (aka the dinghy) wouldn't start or stranded us on a beach. And because of those memories, we really and truly appreciate the things that make our life easier and more comfortable now. We can live without luxuries, we've proven that, but we really don't like camping. We know lots of people who view their time on a boat as a trip and are willing to make sacrifices for a short period of time in order to travel. Our boat, by contrast, is our full-time dwelling (whether we're actively traveling or not), and because it feels like a home, we have been able to make our lifestyle sustainable.

That's not to say we don't miss having the space of a house, or the access to land-based amenities. When we visit grandparents or stay in Jay's grandmother's home in Naples, our kids love to run around in the yard or go to a nearby park, spread into far corners of the house, and take hot baths (admittedly, I too like a candlelit soak from time to time). When we take overland trips, we love to rent houses using Airbnb, for example, and we do not rent dumpy little houses, either. We have stayed in some large and luxurious places in gorgeous locations; whether we're overlooking the Pacific Ocean in Costa Rica, staying in the middle of the Cloud Forest in Panama, looking down from the balcony of a historic Spanish Colonial building in the walled city of Cartagena, or living in the shadow of a volcano in Guatemala, we espouse this view: go big or go home.

And what do we appreciate about these houses? What are the things that we used to take for granted before we moved onto a boat? Acres of kitchen counter space. Walk-in closets. Privacy.

Long, hot showers. Room to sprawl or to disappear with a good book. Garages. Privacy. Gardens. Beds as big as an ocean, which I must swim across to find my husband. Air-conditioning. A place to park a car and unload groceries. And did I mention privacy? It's amazing how freeing it is to be in a room by yourself, thinking your own thoughts, without having to share everything, all the time. On the boat, the only time I'm by myself is when I drop my kayak in the water and paddle off. Otherwise, everyone is within earshot. As an extrovert, that loss of privacy isn't a deal breaker for me, but I imagine it's tough for my introvert husband—no wonder he likes it when I disappear with the kids on field trips!

For nearly all of February of 2018, we had been hemmed in behind the reef in San Andrés, Isla de Colombia. We'd had such a rough passage from Panama that we needed some recovery time in the calm anchorage behind the barrier reef to forget about it and to plan for better weather to continue. When the wind abated and the roaring, foaming, continuous curl of wave that marks the edge of the reef disappeared, conditions became amenable for the day sail to Providencia. We called the port authority as we motored out of the channel one morning and raised sails for the fifty-nautical-mile trip. The ferry, a small power cat, passed us on its way to Providencia midmorning, and again on its way back midafternoon, a reminder that our sailing life is more about the journey than the destination. Otherwise, we could have just forked out the dough and zoomed over to the sister island for lunch.

Instead, we arrived at sunset and stayed for weeks, not hours, in the bay of this idyllic little island, the opposite of San Andrés. Where San Andrés is covered with hotel complexes and buildings practically on top of each other, Providencia is sparsely populated, with low, brightly-painted houses and small private inns. Where San Andrés has an airport with eight flights arriving daily from Bogota, Providencia has only a tiny airstrip, and nowhere to put the one million visitors per year who want to escape mainland Colombia for a vacation in the islands. In San Andrés, you hear Spanish in the market and rarely meet an islander with ancestral ties, but in Providencia, you often hear the lilt of West Indian-accented English and meet few mainland Colombians. San Andrés is mostly flat and developed, Providencia is mountainous and green. We loved it there.

For three glorious weeks, we snorkeled in the clearest water I have ever seen, watched Sam and his new friend Finn sail a little dinghy around the harbor, jumped off the rocks at Morgan's Head, explored the island by golf cart, hiked up the peak, inflated our floating island of fun so the kids could swim and play in the backyard, hung out with Luis and Tachi, shared meals with sailing buddies, watched Luis practice with his band, and toasted beautiful sunsets from the upper deck of *Aqua Lobo*. These are the good things that we will talk about for years to come.

And what about the bad things that never happened to us? For example, hearing about acts of piracy off the coast of Nicaragua and Honduras, we decided that when we sailed north from Providencia we would stay away from the shallow Cayman Banks and not cut the corner to the Bay of Honduras. To be on the safe side, we turned off our lights and AIS transmitter and stayed about a hundred miles offshore; during the two-day sail to Grand Cayman, we saw…nothing. No lights, no small boats, no armed men, not even fishermen. A month later, there were multiple incidences of sailing vessels being pursued by suspicious speed boats, radio channels being jammed so they couldn't call for help, and narrow escapes, so maybe it was a case of being in the right place at the right time. Regardless, we have no pirate stories.

We also have never been struck by lightning, never experienced anything above the rare fifty-knot gust, never seen waves larger than those in the ten- to twelve-foot range, never hit a reef, never sprung a leak at sea, never lost a man overboard, never been entangled in fishing nets, never caught fire, and never come into close contact with man-eating sharks. It is probably dangerous to even talk about these things. Sailors are notoriously superstitious, and just boasting that we've never experienced these hardships means I ought to be frantically knocking on wood. Maybe if we keep sailing, if we someday cross oceans, we'll have opportunities to brag about close calls with these and other dangers. However, we tend to play it pretty safe, obsessively checking the weather before passages, reading other people's stories in books and blogs, and trying to be prepared for anything. We have very few harrowing tales to tell, and for that I am thankful.

That early spring in Providencia was the beginning of some of the nicest cruising we've ever enjoyed. We arrived in Grand Cayman, performed a hassle-free haulout for an insurance survey and made a quick repair, invited my mother-in-law, Mary, for a weeklong visit, signed Sam and Sarah up for an open water SCUBA diving course, did some fantastic freediving, and swam with the famous inhabitants of Stingray City in the North Sound. We rented a car and toured the island, visited the Turtle Center, met the blue iguanas at the Queen Elizabeth II Royal Botanic Park, ate Barbecue at a place the locals recommended, spent a day at the famous Seven Mile Beach, took the boys to the Black Pearl Skate Park (ranked largest in the Western Hemisphere), and stocked up on gourmet items at the nicest—and most expensive—grocery store in the Caribbean. Jay and I got a couple of date nights at posh restaurants, and Eli and I went to see a movie on the big screen in the fancy theater at Camana Bay. It was loads of carefree fun.

We had a gorgeous downwind sail in early May from Grand Cayman southwest to the Bay Islands of Honduras, where we spent three weeks island hopping and enjoying pristine reefs, meeting new friends, hiking to a waterfall, getting a freediving certification from the pros in Utila, and even swimming with whale sharks. If it gets better than that, I can't imagine how. These are the rewarding experiences that make up for discomfort, isolation, difficulty, and character-building life lessons. I hope that these are the happy memories our kids will take with them into adulthood and recount when they talk about their childhoods, even though I know the best stories are the ones that involve storms at sea and all the things that went wrong. Perhaps without rough days at sea, we would not properly appreciate smooth sailing.

# 17

## COURSE CORRECTIONS
### FLEXIBILITY

*June 2018.* Take Two *glides slowly past towering limestone cliffs strung with vines as if decked out for a holiday. Trees grow down from the tops and up from the bottoms, foliage reaching up for light and roots down for water. Between the walls of white rock and green vine, the dark water flows slowly and eddies lazily near the banks, and we find ourselves in midday twilight. Our sixty-eight-foot mast is dwarfed by the 400-foot cliffs, birds swooping and wheeling high above us. We stand on the cabin top, lean against the boom, and watch the new surroundings slide past. A family paddles by in a* cayuco, *two small, shiny-eyed toddlers staring back at us while the mother and father intently feather their paddles. Fishermen stand in the bows of dugout canoes and throw cast nets for bait, or lay nets across the river using plastic soft drink bottles for floats. When we exit the canyon, the river opens out into a wide bay. There are green hills, mist-covered mountains, jungle islands. Along the banks, we begin to see palm-thatched dwellings with docks reaching out from the shallows, reeds obscuring shoreline. White egrets or blue-black cormorants rest with open wings on bare trees that rise out of the*

*water. All is quiet and peaceful, the only sound the rumble of our diesel engines.*

We round a bend and head toward Texan Bay, where we will anchor for the night before motoring the rest of the way to Fronteras, our hurricane season destination on the Rio Dulce, Guatemala. It is early June, and we have just checked in at Livingston, a small Caribbean town on the coast just south of Belize. Surrounded by dense jungle, it can only be reached by water. Polite government officials arrived by lancha, a small motorboat, and sat in our cockpit eating cookies and drinking coffee while we answered questions and they signed and stamped papers. I have enough Spanish after our year in Panama to make light conversation, asking about the weather, the local news, the status of the Volcán de Fuego, which has recently erupted. Overall, it was a surprisingly quick and pleasant experience; we have grown patient living in Central America and expected a long, drawn-out process. Now, as we coast in toward the shore and prepare to drop the anchor, I am struck by the permanence of our decision to come here.

The ocean lies miles behind me, it's long thin line of horizon already a distant memory. We are hemmed in by green mountains, by trees, by river reeds growing in fresh water—an inland hurricane hole where we won't have to worry about storms or wind for at least six months. Some of us are excited by the prospect of staying in a marina, exploring the country by bus (Mayan ruins, mountain villages, Lake Atitlán), learning more Spanish, working on boat projects, and finding a community of boaters again. Others are merely resigned to the logic of the plan but not happy about the extension of our travels for another year.

There's nothing unusual about a ship changing course. One might steer away from another vessel to avoid a collision, to go around shoals, or to adjust for changes in wind and current. A course correction may be as simple as bumping the autopilot a few degrees to port or starboard to allow leeway or to stay inside a channel. Remember when we were sailing off Cape Canaveral and the officials asked all vessels to divert outside their fallout zone for a rocket launch? And the time we were sailing off the coast of Virginia when the U.S. Navy asked us to change course to avoid an area where they were doing exercises with live ammunition? Those were no-brainers. A failure to course-correct can have many negative consequences—running aground, a collision, ugly weather…overspray with stray bullets.

Other times, an entirely new course is set—a waypoint is chosen and a new rhumb line drawn in order to steer toward a different destination. Engine failure, leaks, squalls, a lightning strike, or an ill crew member: all these things can cause a boat to turn about face and seek a safe harbor. The crew of *Take Two* are fair-weather sailors, so we change our minds and turn around all the time. The running joke is that we take two tries to go anywhere. We study the weather meticulously, never depart based on a calendar, and always give ourselves a bailout point.

We know people who sail according to a calendar; sometimes they have good luck and everything miraculously works out. Other times, they get their butts handed to them by Mother Ocean—and when she ain't happy, ain't nobody happy. We may have had such plans when we first started out, but we are no longer those people. Now, when we have an idea about a sailing trip, we discuss it tentatively at first, tiptoeing gently around it, touching it with a stick, then poking at it from every angle (starting with plan A, B, C, and D, and ending with the worst-case scenario), until finally we have beaten it to death. We barely even look at a calendar. If we do manage to set a timeline for departure, it is merely a suggestion, and we overshoot it more often than not. Sailing in less than ideal conditions has taught us to plan carefully around wind and weather and not around holidays or weekends or other people's vacation dates. This means that all plans are subject to change; for type A suburbanites, this adjustment was very difficult.

When I was five years old, I had my whole life planned out. I'd get good grades, go to a good college, be a teacher, get married, have a family, and write a book. Check, check, check, check, check...and check. I'm a very determined person, which is both a blessing and a curse. When I decided that sailing with my husband, teaching my children in a real-world classroom, and going on an adventure were also worthy goals, nothing could stop me. Nothing, that is, except the harsh reality of bad weather and breakage. What happens when a stubborn planner runs up against the brick wall of the elements of nature and the unpredictability of a sailing life? Extreme frustration with a chance of temper tantrums. It isn't pretty.

In order to enjoy the exciting life I dreamed about, I had to learn flexibility—my rigid ideas about how things ought to be had to give way to acceptance of how things are, or the trip would have ended a long time ago. This required a major course correction. To be honest, I have spent my fair share of time crying in frustration, usually in the privacy of my own cabin, but sometimes while walking on a beach, kayaking alone in the mangroves, or scrubbing the galley floor. Fortunately, my even-keeled captain does not seem to suffer from the same illusions of control or bouts of frustration, and models calm acceptance to our crew. Changing course doesn't have to be an emotional event.

Sometimes course corrections come in the form of rude

awakenings—we're sailing along, minding our own business, and we look up at the horizon to see something unexpected in our path. Probably the biggest change we ever faced on *Take Two* was the birth of Rachel in 2011. After we bought our boat, moved aboard with four children, sold our house, and left our first marina, we thought we were on our way around the world. We felt unstoppable. Our kids were young, but no one was in diapers. They had adjusted to life on the boat and were ready for adventure. We were preparing the boat to leave the Florida Keys for the Bahamas and beyond. That's when I discovered I was pregnant.

I called Jay on his way home from a business trip to break the news. I wanted to give him time to think about—and carefully craft—his response, as he drove back to the Keys from the Fort Lauderdale airport. While a baby is always good news in our family (we'd already had a few happy accidents), it would certainly put a damper on our traveling plans. Not to mention where we would even put a newborn; it would mean renovating the boat to make a safe and comfortable space for a baby. Six-year-old Sarah was the happiest about the change; she had been asking for a baby sister for several months, adding the request to her bedtime prayers. We hadn't planned on giving her what she desired…so much for our plans.

Not knowing how adding another family member would impact our ability to live footloose and fancy free, we decided to travel before the birth and figure it out as we went along. When we left, I was three months pregnant. We returned to Florida for the birth of the new baby.

We brought our new baby home to the boat and adjusted to a new rhythm of life. It took us a year to refit the interior, a project that turned into a complete remodel. Add another couple of months to get up the guts to sail away with a baby. And what happened when we finally left? That's when we ran into Tropical Storm Debby and had to turn around after trying to leave Tampa Bay! Let's just say that things rarely go according to our plans, and yet often turn out better. The life that I have now is so much better and more interesting than the safe and predictable life I had planned for myself.

In November 2012, we attempted to exit the channel at the Fort Pierce inlet, the first step in a passage to the Bahamas. We had done our homework and the weather looked pretty good for a Gulf Stream crossing. It was the week before Thanksgiving, and we were planning to spend the holidays in clear water anchored off white sand beaches, instead of sitting at a dock in drizzly, gray weather. Of course, despite our best intentions, exiting the channel was gnarly. The wind and current were having a little argument and taking out their anger on the waves (and us).

The first few minutes of a passage are often the roughest as the waves tend to stack up close to shore. The sea evens out and things do settle down, but first you have to make it out past the rollers. Rachel was a year-and-a-half old and had become accustomed to the boat being calmly tied to a dock. The unpredictable movements of the boat were scary for her, so she crawled up into my lap and we crammed ourselves into a corner of the cockpit. The other kids were already prone on settees. As catamaran owners, we are especially bad at stowing loose objects because we have a basic assumption of stability, so when things started falling and rolling around, I was helplessly pinned down by a frightened toddler wrapped in a blankie. The bell rang as the boat gave a lurch in the waves. I had forgotten to fasten the fridge door shut with duct tape, and I heard things crashing and spilling. I looked sideways at Jay, and he at me. An unspoken conversation passed between us that went something like this:

I raised an eyebrow, which meant, "Do you feel like doing this today?"

He pursed his lips and responded wordlessly, "No, not really." His eyes asked, "Do you think we should turn around?"

I replied, raising both eyebrows for emphasis, "Heck, yeah!"

He acknowledged with a nod, double-checked aloud, "So we're going back, right?" and began to make the necessary course correction.

Unless you have traded sparkling days in an aquamarine sea anchored near remote islands for a damp, cold anchorage in the mangroves and a Thanksgiving trapped in a small cabin with five children, you can't really imagine our chagrin. But this is what it means to be flexible: to bend, to stretch, and to move with changing circumstances. It is the opposite of being stiff: stubborn, brittle, and broken by consequences or disappointment. So we sucked it up, roasted a turkey breast, baked pies, and drank a bottle of wine. And left a week later in much better conditions, with the fridge taped securely.

To this day, we have our own wind-and-wave scale: we might say, "it's a fridge-spilling kind of day," or "it's a bell-ringer." And we aren't afraid of turning around and heading for shelter in that kind of weather. In 2016, while island hopping in the Leeward Islands, we tried to sail to Montserrat from Nevis, but were turned back by unpleasant seas. The officials seemed very understanding when we checked back in, tearing up our paperwork and telling us to come back when we were ready. We also checked out twice from Providencia in 2018, the first being a false alarm. More times than I can count, we have sat in the cockpit, everything ready for a trip, and had a powwow over changing weather or a broken part that prevents a departure. Jay usually sighs with relief ("No pressure, captain!") and I have a good cry ("All that preparation for nothing!"). Then we go about our day. We all handle disappointment in our own way.

Here's another unofficial *Take Two* weather term: Junish, as in, "the weather feels a little Junish." This condition is characterized by overcast skies, afternoon thunderstorms, oppressive heat, alternately calm water and rough chop when a sudden strong wind comes up. It signals a change of season, the beginning of tropical waves and hurricane formation. We were in Belize in early June of 2018, having sailed west from Utila in the Bay Islands of Honduras, slowly making our way north toward Mexico and, ultimately, Florida.

Believe it or not, we were tired of palm-clad islands and isolated anchorages, tired of moving all the time, tired of checking in and out of countries, tired of jury-rigging things instead of fixing them properly because we didn't have access to the right parts and supplies, tired of each other. And we were tired of waiting for the weather to cooperate so we could get out to the much-praised atolls of Belize and the second longest barrier reef in the world. We were puttering around the numerous islands off the coast, wandering aimlessly, not really wanting to stay, but not really wanting to go. And the weather was starting to feel Junish.

About half of the family wanted desperately to get back to Florida, despite the fact that Hurricane Irma had basically destroyed the Florida Keys the previous fall and reminded everyone why coastal Florida is not the place to be during hurricane season. The other half wanted a change of scenery but didn't want to stop traveling just yet. Furthermore, *Take Two* was looking a little worse for wear. It was time for a bottom job, new blue paint for the topsides, and some work on the decks. That meant heading toward a boat yard—either north to Florida or the East Coast of the U.S., or south to Rio Dulce, Guatemala. A paint job in Guatemala would be half the cost but would require some finesse—acquiring supplies, communicating with workers in Spanish, and figuring out where the family would live while the boat was on the hard. We had always been curious about Guatemala in general, and about the Rio in particular, as it is a good hurricane hole in the Caribbean. It was a mere two-day sail to the south, while reaching Florida meant at least another week at sea during the beginning of hurricane season. How could we satisfy all these needs and wants?

Now that our kids are older, we often ask for their input. In our family, everyone gets a say, even when they don't get a vote. Their feelings matter to us (this wasn't their dream, after all), and their wisdom is often appreciated when we're making a decision, as they think of angles we might not. We told everyone to think about it, pray about it, and consider the two options: A) a sail north to return to Florida, where our friends and family were waiting, but we would have to do expensive boat work and have to plan for hurricane season, or B) a sail south to Guatemala to spend a season at a marina in the Rio Dulce, where there would be other boat kids, lower living and repair costs, and a chance to do some overland travel, but would delay our return to the U.S. by about a year. Plan B would also give us a second chance at Belize the next spring and allow for time to stop and enjoy Mexico. Everyone agreed that it made sense to take the boat to Guatemala, even though it might not be their first choice personally. We offered a consolation prize: a trip by plane back to Florida for a visit with friends and family, ostensibly while the boat was getting painted. We hoped it would be a win-win situation.

We found everything we were looking for in Guatemala, and

more: a nice marina (inexpensive by U.S. standards), old friends, new people with whom to practice music, and lots of other boat kids, including an unusually large group of cruising teenagers. Guatemala was a pleasant surprise with its produce stands piled high with gorgeous locally-grown fruits and vegetables, hard-working crew to work on *Take Two* in the boat yard, friendly locals with whom we could practice Spanish, and opportunities to get out in nature—trekking in the jungle, hiking on volcanoes, swimming in waterfalls, ziplining, kayaking on the river, and viewing exotic wildlife in our backyard. It offered opportunities to get involved and give of our time and energy: we volunteered at a kids' camp and a children's home, taught English, and started a language exchange with young people we met along the way. We learned to love Guatemalan food and culture and found a travel company that operated a *bus privado* which we used to explore Mayan ruins, the Spanish colonial town of Antigua, and Guatemala City. And we didn't check the weather once in ten months. It was either rainy, or it wasn't. We received a reprieve from worrying about wind, current, tides, and wave height. Tied to a dock, we didn't have to worry about safe anchorages, power, water, or fuel consumption. We all breathed a sigh of relief.

We kept our promise to the kids and returned to the United States after two-and-a-half years away. We planned an eight-week visit and took our first-ever family road trip. It started with summer lobstering in the Keys at the southern end of US 1 and ended with fall foliage in Acadia National Park at the northern end in Maine. It included stops to visit family and old friends, some of whom we had met on boats in previous cruising seasons. We drove scenic highways, stopped at landmarks and National Parks, ate at diners and fast-food joints, read stories, listened to music, laughed, cried, and fought. It was a typical American Road Trip.

We spent a few days with our friends Joseph and Marina and their boys Owen and Zuber in their Maine farmhouse before heading south again to be with Jay's mom on her seventieth birthday in St. Simons Island on the Georgia shore. We had met them when they lived on their boat, *Little Wing*, and had even been neighbors a few times. We gave ourselves five days to get from Maine to Georgia—plenty of time to go slowly and enjoy the drive, and maybe even stop at Gettysburg or another landmark on the way. It was early October and the trees were showing off all their best colors, throwing red, yellow, and orange confetti with every gust of wind. We climbed into the family Suburban and waved goodbye to our friends.

And then came the pivotal moment when I knew that living on the boat had really changed me from Someone Who Freaks Out to Someone Who Goes with the Flow. One minute we were zooming south on the highway, and the next, we were standing on the side of the road with a car that appeared to be in its death throes. We had pulled over in Brunswick, Maine after the Suburban made a sound like nails on a chalkboard and found a couple of helpful mechanics who sounded just like Car Talk's *Click and Clack*, the Tappet Brothers, and who gave us the bad news. We ended up at the Chevy dealer, ordering a new transmission. The catch? It was Friday afternoon, and they couldn't order parts until Monday morning. They wouldn't be delivered and installed until Tuesday afternoon. We were due in Georgia on Wednesday for Mom's birthday. This was not what we had planned.

Here's the important thing: I didn't rant or throw a temper tantrum or blame anyone. I didn't weep, kick the car, bite someone's head off, or even shake my head bitterly. I didn't laugh like a lunatic, but I did find the situation mildly amusing. I thought, *How fitting! We've gone on an old-fashioned road trip and even this trip has been foiled by breakage!* Everyone ought to stand on the roadside by a disabled vehicle at least once in their lives, wondering what comes next. We reassured the kids that everything would be fine, that we were now having "an adventure." I admit that I went so far as to think, *this will make a good blog post.* I was already imagining how this would be something to laugh at someday.

If that's not a course correction, I don't know what is. It was not the same perspective that I had ten years ago. Remember when I forgot our overnight bag at the house when we went down to the boat for the weekend? When we got to the marina and unpacked the car, I threw a little hissy fit in the parking lot. The rest of the family left me standing there in a literal and figurative drizzle, probably because they were embarrassed, and because they didn't want to get the only clothes they had wet. Stranded roadside in Maine, I was no longer that woman. (Not that I don't throw hissy fits, only that it takes a lot more than it used to!)

We called Joseph and Marina from the parking lot of the Chevy dealer, where we were unpacking the back of our car. Our dear friends, who loved us like family, bailed us out. Since there wasn't even a rental car big enough for our family available last-minute in that small town, Joseph drove an hour to pick us up and drive us back to their house. Marina welcomed us back into their home with open arms, and our kids spent a few more days picking apples, stacking wood, and playing games with their friends. Instead of a rushed visit, we enjoyed leisurely afternoons walking in the fall-clad forest or along the rocky shore in a crisp breeze, afterward sharing tea, music, laughter, and deep conversation by the woodstove. And the icing on the cake? The good people at the Chevy dealership fixed our Suburban as quickly as possible, knowing that it was important for us to keep our promise to Jay's mom, and we drove twenty hours straight from Brunswick, Maine to Brunswick, Georgia, and made it to the birthday gathering in time.

What I have come to realize living on our boat is that all of life can be seen as an adventure, but it requires a mental shift, an openness to spontaneity and change, and a willingness to participate in a Plan bigger than our own. It's a course correction in the largest sense—the ability to take a detour or change routes quickly and calmly when the need arises. Looking back, we can follow the trail of ifs—if we hadn't missed that weather window, if we hadn't anchored where we did, if we hadn't broken that part and stayed in port...or gone here, or done that, or met this person...then we wouldn't be where we are right now. Life is a series of happy mishaps, of small changes in direction that result in arrival at a different destination. Small course corrections can mean big changes in trajectory, the navigational equivalent of the butterfly effect.

The good and the bad, the interruptions and the detours, all get woven into a bigger story. As we avoid obstacles, or steer toward different opportunities, we sense God's leading and purpose for our lives. Our contentment depends upon whether we resist or cooperate with a change in plans. One of my favorite verses from the Bible comes from Paul's letter to the Romans: "And we know that in all things God works for the good of those who love him, who have been called according to His purpose." I used to hope that that was true but looking back at more than a decade of living on a boat, I know it is.

## 18

## SAFE HARBOR
### LETTING GO

*April 2019. We just met Eli and Aaron near the fountain at the park, to retrieve them after their overnight hike up the Acatenango Volcano. Jay and I are now sitting with two dirty, bleeding, and happy teenage boys at a barbecue restaurant in Antigua, Guatemala, waiting for cold beverages. They seem like strangers, or old friends I haven't seen in a long time. Because we homeschool our kids and spend so much time together, sometimes I forget to look at them. Yes, I see them every day—but unless we spend time apart, I don't really notice the gradual changes. And there has been a subtle but important transformation here, like a rite of passage. Without a First Buffalo Hunt or similar ceremony, our culture fosters an extended childhood and adolescence, making it hard for a boy to know when he's become a man. But make no mistake: these two are strapping, handsome young men, not the boys with whom we set sail several years ago.*

*Early yesterday morning, Jay and I walked with the two of them to the Central Park by the cathedral to meet the guide and see them off. They were wearing boots and carrying backpacks full of water and warm clothes for the freezing temperatures they would find at the summit. I hugged them and wished them well, then Jay and I went* out for coffee while the other three kids slept in back at the rental house. It is nice to just wander around the colonial city, just two people instead of a troop. I can imagine a second honeymoon experience when the kids are gone. It will be a consolation to have that core relationship brought back to the forefront after the hectic years of raising children together. We have worked hard to keep the romance alive using stolen moments like these, so that we'll have something to come back to.

Last night, when the five of us sat down to eat dinner it was eerily, and some might even argue pleasantly, quiet. While Eli and Aaron camped overnight on Acatenango, the volcano neighboring the actively erupting Volcán de Fuego, we got a small taste for the changing family dynamics. It's something we've noticed before: any time you add or take away a person from the family group, the chemistry changes. Jay looked positively giddy, already imagining the peace and quiet of a house with fewer children, but I'm having mixed feelings.

If a mother does her task well, she works herself out of a job. I know this, and I have been aware of the metamorphosis of my teenagers, but I am not entirely emotionally prepared to let go. It is true that I am always excited for a kid who jumps on an opportunity to be independent—to ride a public bus, to take the dinghy ashore, to go on a hike that's too challenging for the larger group, to work or meet friends without the ball-and-chain of the family. I remember how fun it was to be free of my family for a few hours when I was a teenager. But I never considered how it made my mother feel to not be needed. And now I'm the mother, and soon my own children will have flown the coop. It's part of the cycle which J. M. Barrie describes in the last line of Peter Pan, after Wendy has grown and sent her daughter to be Peter's mother: "thus it will go on, so long as children are gay and innocent and heartless."

Nothing can describe the relief one feels after coming into a calm anchorage after a rough day at sea. I imagine it is akin to finding an oasis after crossing a desert or seeing the lights of a cabin after being lost in the dark woods. To drop anchor in calm water, out of the wind and waves, especially along a familiar shoreline or in a bay we've called home: this is what it means to find a safe harbor. It is the promise of a warm meal, rest for the weary crew, freedom from worry, and a good night's sleep. It may mean finding fellow travelers and rejoining a community, the chance to meet friends old and new. Properly appreciating such a welcoming place, of course, means having left a safe harbor somewhere else, and having taken risks, had adventures, and experienced discomfort or danger.

At the same time, the safe harbor isn't somewhere you want to stay too long. Comfort quickly becomes complacency, and complacency, apathy and boredom. One wants to avoid the Velcro ports. As John Augustus Shedd has said, "a ship in harbor is safe, but that is not what ships are built for." Safe harbors are great for taking a breather, preparing for another voyage, repairing a boat, provisioning, and receiving landlubber visitors before setting sail again. They are jumping-off places, springboards, and respites for catching your breath before the next leg of a journey.

At the end of every season, there is a migration which mimics one in nature—sailors grow restless like geese in late summer. Boats that spent a season island hopping, coastal cruising, or anchoring for a while in a beautiful place, depart for cooler or warmer climes. They return to a home port, sail off to hurricane holes or inland waterways, or haul out for repairs. There's just a feeling in the air when it's time to move on. Even those of us who live on land are familiar with the alternating feelings of relief at coming home and readiness to leave. For everything there is a season and for every voyage there is a starting point and a destination. For every destination, there is another voyage calling and another departure to be made. It's always bittersweet because for every reunion there is a goodbye, for every welcoming home port, there is a longing for distant shores. In a nutshell, this tension between contentment and aspiration, and the ability to hold people and places lightly is what defines a nomadic life.

The task of letting go begins with the cutting of the umbilical cord—I remember the joy of being unburdened at childbirth but also the sense of emptiness when my small companion of nine months was suddenly on the outside. The realization that my whole task was to bring this helpless, squirming thing to adulthood and independence hit me like a freight train. How is it even possible? On the other hand, in the animal world, eighteen to twenty years of rearing offspring is unheard of—our closest cousins, the chimpanzees, raise their young to independence in about ten years. Our marine mammalian cousins, Dolphins, are born swimming and reach maturity around age eight.

Not so with humans; we're slow learners. But one day at a time, one step, one tumble, one bike ride, one swim, one tree climb, one dive at a time—our babies grow up and learn independence, too. On *Take Two*, we try to foster autonomy and responsibility, even while homeschooling in a self-contained environment. We have often had access to small sailboats and kayaks, where our children could get out and about on the water and learn the ways of the wind and waves. Eli took me sailing in an Opti, a small sailing dinghy, when he was eight years old at the end of summer camp at a yacht club. By age nine, Aaron had his boating safety I.D. card, giving him the ability to run around in the dinghy by himself. At the same age, Sarah could go kayaking by herself, and she had access to the galley and became quite the baker. Sam could catch and clean enough fish to feed our family. And Rachel, at eight, could already make homemade tortillas by herself.

By age twelve or thirteen, a crew member is ready to take a night watch while we're on a passage, responsible not just for themselves but for the safety of the boat and the whole family. And Jay and I can dinghy in for date night now and leave one of our teenagers in charge of the boat and crew without worrying about mutiny, fire, or sinking. Our kids are rock climbers, cliff jumpers, free divers, spearfishers, and sailors; even though it is scary, we let them learn out in the wild where the risks are real. I would argue that the alternative, a safe life lived indoors, carries a different kind of danger.

The work-study part of our homeschool that began while we lived in Fort Pierce in 2015 helped our kids explore the world of adulthood in an engaging and meaningful way. Aaron went to work once a week with our friend Ben, where he learned all about boat repair—and how to climb into some hard-to-reach places. He spent time in the shop with the old salts, worked on engines, swept and organized, got his hands dirty, went out to lunch, and got a taste of the workday world. Eli's flying lessons and friendship with his instructor gave him insight into what a career in aviation might look like. Sarah spent many mornings with her riding instructor, Stacy, who showed her what it was like behind the scenes at the stables. She mucked out stalls, fed and watered animals, brushed and bathed horses, and learned about running a barn and a business, in addition to spending time on horseback. Sam was learning to kickbox along with his older siblings, and the instructor, Jim, a fellow boater, would occasionally take him fishing, where he learned tricks of the trade. It was our goal that a kid who graduated our homeschool would have knowledge of both academic subjects and practical life skills.

In Panama in 2017, Eli and Aaron started volunteering in the small-boat repair shop at Agua Dulce Marina for our friend Bobby. Organizing and doing odd jobs quickly morphed into working part-time for pay; Eli and Aaron would do school in the morning and then take the dinghy to work in the afternoons, where they disassembled pangas to prep them for fiberglass work and paint or reassembled them after the work was done. The kids all have experience doing odd jobs on other people's boats—cleaning dinghies, going to the top of a mast, polishing, fetching, carrying and loading, and generally lending an extra hand.

The volunteer work in Guatemala provided valuable experiences; some of our kids and their boat friends worked with kids at a summer camp doing sports and games and later helped at an orphanage, carrying rocks, digging a trench where a boat would someday be docked, clearing and cleaning around the property, and sometimes just hanging out with little kids. In addition to learning the value and reward of hard work and altruism, they also learned how to speak another language, navigate a new culture, build relationships, and get around in another country.

Eli and Aaron had stretched both their confidence and their independence when they came back from their hike to the summit of Acatenango, where they witnessed a marvel of nature —an eruption spewing red fire and liquid rock into the night sky, rumbling and shaking so they couldn't sleep as they huddled in their tent perched on the side of the neighboring volcano. On the way back down, they'd both taken a tumble on the loose igneous rock as they trotted down the trail. Was I upset that Eli had torn his jeans? Or that Aaron was bleeding? Heck, no! That's what we sent them up there to do: to have a good time, to see something amazing, to make a memory together, and to take risks. We know it costs something, but what is purchased is priceless: the confidence to go out into this big world and live life to the fullest. It's what every parent wants for their child. It's worth letting go to see your kids become who they were meant to be. I feel grateful for the boat life that allowed us to be a part of that transformation. In essence, that's why we set sail in the first place: to let go of our inhibitions and see what a life of adventure could teach us about ourselves and the world we live in.

Music is another avenue to independence—Eli and Aaron played their first open mic night at a cruisers' gathering at the St. Francis hotel and restaurant in the Bahamas in 2016, and since then, there have been a lot of opportunities to practice and play with other musicians in diverse settings. While the boys were working at Agua Dulce, Sarah was riding along with them to work to meet with her friend Ellie to practice guitar and ukulele. At a Shelter Bay open mic night, Sam joined the act and played drums. In Isla Mujeres, Mexico, Aaron met up with a guitar player who gave him a few lessons and handed him the microphone for a few songs onstage at a local bar. All of this life experience is what we had hoped for when we bought the boat. Everything we do—running and repairing the boat, making passages, preparing meals, exploring islands, taking classes, meeting new people, hiking in the wilderness—should be expanding our kids' horizons and preparing them in some way for a full and happy life out on their own. And in the process, we grown-ups learn something too.

I was walking on the beach with Rachel along the shore of West End in Roatán, Honduras one afternoon. It was early May 2018; the water was calm and clear, and we had been anchored for about a week just inside the reef in a large sea-grass bed. It was a beautiful anchorage—great snorkeling and diving in our own backyard and easy access to a shore sprinkled with boutique hotels, dive shops, and restaurants. In winter, this anchorage can be untenable; cold fronts bring wind and swell which drive the boats to the sheltered southern bays of the island. But in May, no one is worried about the weather...unless a sudden thunderstorm pops up.

When we bought our anchor—an eighty-pound Manson with a roll bar for easy re-setting—people on our dock laughed at us. What on earth did we need such a big anchor for? I guess they had never stayed awake on a windy night watching the chartplotter to see if their boat was dragging toward the rocks. But we had anchored in the Bahamas, where the boat can swing 180 degrees with the shift of the tides, and we had dragged plenty of times even in protected anchorages when squalls came up. We never slept as well as we did after buying that anchor. In the seven years we'd owned it, the boat had never dragged...until that afternoon in Roatán, when Rachel and I watched helplessly from the shelter of a nearby dive shop as *Take Two* swung into the approaching storm and then slowly changed position in relation to the other boats.

In the sudden pouring rain, I could see our crew scramble to close hatches and then realize they were dragging. I am the crew member primarily responsible for setting the anchor. Sarah is often my right-hand man, helping to find a sandy spot in which to drop the anchor, deploying the chain, and fastening the bridles. I knew she could reanchor the boat if I wasn't there. Jay was at home working, so I wasn't worried. He had trained Aaron at the helm, and I watched with pride as my two big kids worked together to pick up the chain with the windlass, drive to a new spot, and reset the anchor. Twice. What they couldn't see was the clump of sea grass on the spade that kept the anchor from resetting properly. Eli is normally our anchor diver and checks to see that it's dug in, but the weather made that impossible. After the storm passed, we cleaned the clump of grass and mud from the anchor and resettled a little farther from our neighbors, but for me it was a demonstration of competence and responsibility—what had started as a boatful of children was now a vessel with capable crew.

We left Guatemala in April of 2019, motoring back out the Rio Dulce between those magnificent limestone cliffs, and pointed the bows northwards at last to begin the final leg of our Caribbean Circle. We gave Belize another try, where we were able to get a glimpse of the barrier reef (though the weather still did not cooperate to make possible a trip out to the atolls.) We hung out in Caye Caulker during the girls' May birthdays, meeting up with friends Royce and Jennifer on *Cerca Trova*—a couple we had met in Florida when Rachel was a newborn. I had the privilege of watching Sarah and her Dad sail a vintage Hobie 16 catamaran back and forth across the bay. I remember thinking to myself, *who is that gorgeous blonde woman sitting next to my husband?* At five-feet-eight-inches, she looks over the top of my head—she's long-legged and beautiful—though she still possesses the combination of tomboy-princess that she had as a little girl.

When we reached Isla Mujeres, Mexico, we were a mere three-day sail from Florida. It was June, and though it was beginning to get hot, there was a nice breeze in the anchorage and the water was cool, aquamarine perfection. We were anchored near good friends on *Rothim*, a boat from Israel with three girls aboard, Naomi, Adi, and Zoe, and *Dreamcatcher II*, a boat from South Africa with a teenage son, Deon, aboard. The kids would swim or dinghy over every day after school and jump off *Take Two*'s high dive and play king of the island on our inflatable raft, dubbed the Floating Island of Fun.

With teenage friends to run around with, our big kids had opportunities to do lots of independent things: Eli went scuba diving with Deon, Aaron took guitar lessons with Adi, Sarah went over to the neighbors' to play music with Adi and Naomi or watch movies. Everyone took turns with Naomi's windsurfer. I could send a couple of kids to the store, or to drop off or pick up laundry. We went on a field trip with Joaquin, a friend from Argentina we had met on a boat the year before, who helped organize transportation and entrance to cenotes in the Yucatán. Even Sam and Rachel, the youngest members of our crew, are fearless and independent—they jumped from the edge of a closed cenote—a twenty-five-foot drop into the dark and ice cold water of a subterranean lake. I took photos from below and watched with admiration, and a little envy—I'm generally more cautious than my children.

Before leaving Mexico, we had a potluck on *Take Two* with the kid boats who were headed back to Guatemala for hurricane season. It ended with a magical midnight swim in the light of the full moon—true to his nature, our firstborn broke the ice, launching himself out into the dark and making a splash. The cruising life creates an interesting kind of independence in a kid: at seventeen, Eli had never owned a phone, driven a car, or taken a girl out on a date, but he could free dive to seventy feet, spearfish for dinner, play bass on stage in front of a crowd, write a novel, do a full day's work, and organize a role-playing game for a group of friends. He had leadership skills and confidence in some areas but would probably feel like a fish out of water in his own culture. The same could be said for Aaron and Sarah, who, at sixteen and fifteen, respectively, couldn't wait to get back to Florida but didn't know how they would fit in with their peers. We were about to find out.

. . .

Here's what letting go really means: releasing our expectations, setting our dreams aside—at least for a time, and sacrificing ourselves for someone else. And it is not easy. Jay and I did not want to leave Central America. I had begun a relationship with two wonderful young people in Guatemala through an English class I was teaching, and I hated to leave them; Wendel and Vivian had become like family. Aaron was meeting regularly with Wendel for a language exchange. The rest of us were also beginning to feel comfortable using Spanish on a daily basis, and we were afraid of losing that newfound proficiency. To be honest, going back to the United States was a daunting prospect; the cost of living would go up significantly, and we'd be participating in a culture with which we really don't identify anymore. Sam and Rachel love the cruising life and going back would mean sacrificing boat travel for a while. On the other hand, the three oldest kids were feeling limited and isolated by our lifestyle. They had an awareness that a life elsewhere was passing them by, and, of course, they were itching to be free. Spending most of your time with your family in a small space is not a teenager's idea of freedom.

So we sailed back to Florida. Or, motored, rather: we drove over placid seas for three days and nights. Instead of choosing sailing weather, which would mean a rough last passage, possibly pitting the wind against the current, we used windless conditions to hop on the conveyor belt of the Gulf Stream, which swept us north and east from Mexico, past Cuba, and back to the Florida Keys. We had briefly discussed the possibility of sailing in a more northerly direction, towards Tampa Bay, but everyone agreed that we should close the geographic circle we had made of the Caribbean and return to the coordinates from which we had left: Marathon, Florida. We remembered an early lesson from our boating life: the way to keep from being overwhelmed is to take baby steps. So we baby stepped our way back to civilization, motoring over glassy seas that reflected the summer sky in its mirrored surface, and tied our bridles to a mooring in the lee of a familiar island in time to celebrate the Fourth of July with old friends. We had returned to the safe harbor, in every sense of the word.

Letting go also meant teaching three teenagers to drive. Immediately, we could see the learning curve would be steep. Within a few weeks of returning to the United States, we had three new drivers practicing maneuvers in our old Chevy Suburban…and I had foolishly thought it was hard to have three toddlers at the same time! But Marathon is a small town on an island, and it felt like a safe place to begin the process. Just as we had weened ourselves away from the land life when we cruised there ten years before, we were now weening our children from the cruising life so they could reconnect with a community onshore. With everything changing so fast, it was comforting to have the stability of good friendships to absorb some of the stress and provide comic relief. Other homeschool families we knew there were also figuring out how to prepare their teens for graduation and the jump to college and work, so we had moral support. We decided to stay until we got our feet firmly beneath us.

It was a season of firsts: Sarah sailed as crew on a Hobie in her first regatta, held at the Marathon Yacht Club. Eli, Aaron, and Sarah took a college entrance exam, their first standardized test, which allowed them to sign up for their first community college classes as part of Florida's free dual-enrollment program. Eli took his girlfriend—a girl he'd known since they were nine—out on their first date. Eli and Aaron took a Greyhound bus to Jacksonville, where they spent a week with Andrew (the captain of

*Abby Singer*) working on a carpentry job, their first forty-hour work week. The teenagers got their first cellphones and laptops—ones they didn't have to share. Sam and Rachel joined recreational basketball teams at the community park, the first time we've ever committed to staying around long enough to participate in organized sports.

We dove head first into the shallow pool of busyness—trying on the landlubber life like we were on a new adventure. It was hard to come from a life where we set the boundaries into one with schedules, rules, uniforms, and expectations; we had a bit of reverse culture shock. But after our triumphant return with all flags flying, and a period of tearfulness as I realized it was over, even I could see that it was time to let everyone chase their own dreams. I took on the role of taxi driver the way I had washed cloth diapers: it was just a new phase in my mothering career. The best part was being able to accomplish these goals while continuing to live aboard on a mooring at the city marina; what a blessing it was to come back over the water as the sun set on another busy day, to see starlight through my hatch as I fell asleep, and to wake to the call of a rooster or red-winged blackbird. These rhythms suffused our chaotic life with peace.

Jay began to build a business he had started the year before, one that might set us free from the time-is-money system and let us travel more in the years to come. I wanted to focus on my writing, and I found a friend with whom I could share that dream and meet weekly to be mutually accountable for progress. Eli wanted to graduate from our homeschool and figure out what to do with his life, including, but not limited to, moving off the boat and getting his pilot's license. Aaron met Tony, an airplane mechanic, at a church he had independently started visiting, and he was offered a job at the hangar twice a week, where he could begin to learn aircraft maintenance and repair. Sarah, our introvert, suddenly blossomed in the positive social environment, taking art and dance classes, joining a Spanish club, and even playing piano in church one Sunday. She was happier than I had seen her in a long time, surrounded by a lovable group of old friends. Sam, who would have rather been out sailing and fishing, reacclimated to living near shore and got his boater safety card so he could take the dinghy ashore to go to the skate park and to play basketball. Even Rachel, who has the tough role of youngest kid in a big family, seemed happy in a group of her peers, and we found a friend with whom we could partner for homeschool one morning a week. It felt good to be back, even though we all missed the freedom of sailing from island to island. In a sense, the stability provided by the safe harbor became a launchpad for a new kind of journey.

In August, Jay and I celebrated our twenty-second wedding anniversary, which marked twenty years since we bought the house with the white picket fence, and ten since we moved aboard *Take Two*. We went back to Naples to visit family we hadn't seen in a long time, and to revisit old haunts. We stopped for ice cream at our Ben & Jerry's and threw a penny in our fountain as we drove by, just like we had done when we were seventeen. It struck me that night: our oldest child is the same age as we were the year we fell in love. It had all come full circle.

Sometimes life goes by in a blur, childhood, college, or the sleep-deprived years of early parenthood, for instance, but there are moments of diamond clarity—mental snapshots we collect that color who we are and how we move through the collage of daily life. Change of scenery, travel, joy, and even hardship can slow the stream of time, mark a moment as special and even create a lens for seeing the future. Whatever happens going forward, the memories we made with our family on the boat are ours to keep. A sunset is not just a sunset to me: every time the sun sinks below the horizon, it contains the blaze of light on the ocean, a green flash as the last rays bend through water, the red glow on the faces of my children as we turn toward the western sky from the top of a Mayan pyramid, the light that fades behind a volcanic mountain peak and the fireflies that blink on with the stars. Something as simple as a cup of coffee now contains not only the highland taste of Colombia, Panama, Costa Rica, or Guatemala, but the place itself—waterfalls, cloud forests, volcanoes, and ancient walled cities. I can close my eyes and go there in one sip.

The single most important thing we learned on the boat was how to let go: of plans, expectations, and fears, and to embrace life in all its beauty and ugliness, hope and sadness, joy and longing. We weighed the risks of leaving the safe harbor, and despite what it cost us, exited the channel and braved the ocean, so that when we returned, what we had gained more than compensated us. Just like it said on our blog, we bought the boat to slowly go broke while teaching our children about the world and having a great time. We came back to the safe harbor with less money in the bank, maybe, but with a wealth of life lessons, friendships, experiences, new perspectives, and happy memories. Living aboard and traveling changed us irrevocably; we may have returned to the same place, but we are not the same people.

## EPILOGUE: IN THE OFFING
### NEW DREAMS

*September 2019. I am looking out over the Rio Dulce from the third-floor balcony of my apartment at Mar Marina. My eyes wander to the slip at Nana Juana where* Take Two *used to be docked...but she isn't there. It is odd to arrive by land in a place I know by water; it makes me feel homeless. I have returned to Guatemala to help my friend Hagit with the birth of her fourth baby, a little boy named Cayo who was born this morning in the wee hours. I'm just waking up after a long nap following a long night.*

*We've had a surreal experience, this unlikely pair of women. I am asking myself,* How did an American find herself sitting next to an Israeli in labor while hurtling down the road in a beat-up Japanese minibus with a crazed Guatemalan driver? *It sounds so much like the beginning of a joke that it made me laugh hysterically even in the middle of the ordeal, which broke the tension and fear—I really thought we might die on that road between Morales and Fronteras. My hysterics were contagious, and soon the two of us were laughing so hard we were crying at the Absurdity of Life.*

*I met Hagit when her family sailed into the Rio last December on* Rothim: *Peter, from Slovakia, Hagit from Israel, and their three girls, teenagers Naomi and Adi, and their little sister Zoe, around Rachel's age. We made an instant connection: I knew immediately that we would become good friends. What I did not know is that I would also be an adopted sister, knitted by the experience of bringing a new life into the world. I came in the stead of her biological sister and mother, who could not come from Israel—it is they who have offered me this stunning view of the Rio, paying for my apartment and giving me a quiet space to which I can retreat after hanging out with the family all day. And my Spanish has made communicating with Doctora Ruth much easier, so I filled the role of translator as well. I had the privilege of holding Hagit's hand during labor and helping with the baby afterward. But though she may feel grateful for these gifts, it is I who have received the greater gift: a two-week mom vacation. I have not been away from my family for more than a few days in eighteen years.*

*In my free time, between helping cook or doing laundry or hanging out with the girls (whom I love like nieces), I have read a good book, written a song, met with my English students, Wendel and Vivian, visited the orphanage where we used to volunteer, played ukulele with my Brazilian friend Ana on* Coragem, *played dominoes with my competitive South African friend Darelle on* Dreamcatcher II, *and reunited with other friends from our season in Guatemala, Rudolph and Elisa on* Tulum III. *This trip is a gift I didn't even know I needed. In the middle of the new life I'm carving out in the United States—one in which I am willingly running from one place to another dropping off and picking up kids as we fully engage in our new environment, and meet the demands of a rigorous school and social calendar—I am suddenly reminded of who I am outside my roles as wife, mother, sister, daughter, friend, and teacher. Just waking up in the quiet apartment, by myself, and having coffee on the balcony, taking time to write without interruption, reading a book for pleasure, walking alone down the crowded main street in*

*Fronteras:* these are simple pleasures that I have forgotten. It took a few days, but I have become comfortable in my own skin again. Of course I miss Jay and the kids! They are the system in which I find my orbit—but at the moment, I am feeling the weightless joy of freedom.

I knew that coming back to Florida would be hard, but I never expected that I would grieve the end of the voyage. It was like having postpartum depression—a deep sadness accompanied the sense of accomplishment. But with all this time to myself, I am beginning to contemplate, for the first time since we bought Take Two, the possibilities the next season of my life might hold. With our kids becoming more self-sufficient, it may be that I get to travel more, not less—little trips to visit friends in exotic places—the crew of Katta 3 for Swedish Midsummer? Jayne from Delphinus at her bed & breakfast on the coast of England? Wendy from Water Lily in Alaska? Kimberly from Ally Cat in Mattapoisett? The crew of Jalepeño in Utah? Who knows? The world is smaller because of the friendships we gained while we were cruising.

---

When you're standing on the seashore and you look off into the distance, what you see between you and the horizon is referred to as "the offing." In the nineteenth century, sailors' wives looked for ships due in port from widow's walks, waiting expectantly for husbands and sons and brothers. From the foredeck or the crow's nest of a sailing ship, the lookout would search the water ahead for obstacles, or keep an eye on the horizon for the welcome sight of land. Metaphorically speaking, what's in the offing is what you can see up ahead: expectations and goals for the near future.

In August of 2017, Jay and I celebrated our twentieth anniversary at Red Frog Resort and Marina in Bocas del

Toro. The kids were old enough to stay on the boat in the marina alone for a couple of days while we rented a beach-front house on the other side of the island. We may have been only fifteen minutes' walk away, but we might as well have gone to the other side of the planet. It was so fun, reminiscent of our honeymoon in Mexico. We went out one night, holding hands as we walked on the path through the twilit jungle to the restaurant on the island. Another night we cooked a gourmet meal and watched an old favorite, *Casablanca*. We slept in, we went for long walks on the beach, we swam, talked, and just spent time together.

The task of child rearing and providing for a family's needs is so all-consuming that it's easy to get swallowed up by it. Add travel planning, homeschooling, and boat repair, and it could be a recipe to kill romance altogether. We have always made date night a priority and tried to keep something simmering on the back burner, so there's never been a question of whether or not we still love each other, but it is so reassuring to have uninterrupted time to remember what we still like about each other.

And we've started to imagine the next phase of life. Since returning from the Caribbean, we started asking, "What if?" again. Like, "What if we gutted the interior of *Take Two* and rebuilt the galley and cockpit to make it suitable for sunset charters or a floating gourmet restaurant? What if we traveled seasonally, leaving the boat somewhere safe for hurricane season, seeing a different part of the world by land or air, and only cruising in the winters? What if Jay's new business was wildly successful and created enough passive income that we wouldn't have to worry about billable hours ever again? What if we crossed the Atlantic and cruised in Greece like we dreamed as teenagers? What if we bought a little piece of property somewhere to have a land base? What if I went back to teaching English as a second language—outside the United States? What if we bought an RV and drove to all the National Parks? What if I could focus on my writing and make it a profession? What if we keep *Take Two* and take our grandchildren sailing?"

We probably won't do all of these things, of course—we'll have to choose. But they are good questions to ask, and a direction will present itself if we remain open to suggestions and continue to communicate and share our dreams. We've got about a decade of child rearing left, so there's plenty of time to travel with Sam and Rachel before we have to think about the empty nest, but if you look at how long it took for us to sail away, it's probably wise to have ideas about the next phase of life.

At the bottom of all these questions is an assumption: we will not be going back to suburbia any time soon. We have plotted waypoints and will stay the course. The life lessons we learned in the process of going sailing with our family, of using our productive years to travel instead of saving and investing, or instead of buying stuff and building a bigger house in which to store it all, means that our unconventional route will take us to a different destination. We met a lovely retired couple, Charlotte and James, on a trawler called *Pegasus*, in Grenada. They are Scottish, and Charlotte shared her thoughts on life and gave me wise advice in her adorable brogue as we took long walks around our end of the island. She started me thinking about what it might be like to be two wanderers in our golden years—and if we end up anything like the two of them, I will count it as a blessing.

Not everyone would be happy living the way we have. We chose the road less traveled, and it is not paved. What has made it sustainable is our ability to set our own pace: to stay if we feel like staying, to go when we need a change of pace or scenery, to shift between fixing the boat, enjoying the boat, and working to pay for everything, and to constantly readjust expectations. As a goal-oriented person, it might be easy for me to focus on failure; for example, when we were young, we said we would circumnavigate the globe someday, but we haven't done it yet, and we may never do it. But our early dreams didn't involve five other people, who didn't ask to go to sea. And we didn't know what long passages would be like, or how much it would cost in time, money, and energy. Just looking at the Pacific Ocean on a map or globe takes my breath away. I mean, three thousand miles is a long way—especially at a speed of seven miles per hour—and that's only to the first place you can make landfall.

Instead, we've made this journey a pleasure cruise and not a race. After a youth spent moving place to place, my ten years aboard *Take Two* represent the longest time I've ever lived somewhere. The boat has provided both contentment as well as a challenge for over a decade. My wanderlust and need for stability are simultaneously satisfied; the cozy home stays the same even as the scenery outside changes day by day. When traveling with five kids felt overwhelming, we were able to return to a safe harbor—familiar places where we have nurtured long-term relationships and were nurtured in return.

Living on *Take Two* has taught us a lot about life and what it's for: growing, learning, loving, and never giving up. We thank God for all the mishaps, trials, mistakes, and storms at sea, and for all the beauty, joy, companionship, and pleasant surprises that together have made our life afloat worth all the trouble. Whatever it is that lies in the offing, we will carry these memories and lessons with us into the adventures ahead.

# GLOSSARY OF NAUTICAL TERMS
## FOR LANDLUBBERS

This book is written for landlubbers by a former landlubber, so I tried to keep the nautical jargon to a minimum. But instead of shouting and pointing, "Move the thing! And that other thing!" I really try to use the correct term. It took me ten years of listening to nautical trivia on morning VHF radio nets to learn the correct names for some of the obscure doodads and thingamajigs on a boat. This is by no means an exhaustive list, but you will find the most basic words and their usages here.

- **Anchors aweigh:** A traditional seafaring term meaning to weigh or lift the anchor off the bottom and prepare for a voyage. I used to think this was anchors away! which, I suppose, would work in the opposite scenario, if you were dropping anchor, or if you were done traveling and wanted to tidy up and secure the anchors. The real sailors are laughing at me as I speak.

- **Batten the hatches:** Archaic term meaning to use

wooden slats to fasten deck hatches so that water stays outside the boat instead of coming in. This is a useful term even now, though we don't use wooden slats, but turning handles that lock hatches to prevent leaks. We use it all the time, or shorten it to "Hatches!" as we scramble around during a sudden thunderstorm to close cabin windows and deck hatches. Also refers to general preparation for heavy weather. It can even be used metaphorically!

- **BEAM REACH:** A boat sailing with the wind on its beam, or side. A close reach is when the wind is coming from the front, either to the right or left of the bow. A broad reach is when the wind is coming from behind, either to the left or the right of the stern. Sailing downwind, or running, is when the wind is from directly behind. In irons is when the wind is coming from directly ahead, preventing any forward movement. One might head directly upwind on purpose, like when raising the mainsail, or preparing to anchor or pick up a mooring.

- **BIGHT:** a curve in a coastline…almost like a giant bite taken out of the land, indicating a wide, shallow bay. Other words for bay include sound, neck, straight, lagoon, cove, inlet, harbor, basin, etc.

- **Boom:** The horizontal beam loosely joined to the mast that holds the bottom edge, or foot, of the mainsail. It swings when the boat tacks or jibes so that the sail can fill with wind. Also, the sound it makes when it hits your head if you are not paying attention during said tack or jibe.

- **Bridle:** A line from each bow to the anchor chain or mooring used to stabilize the boat when anchored and take pressure off the windlass (a device which hoists the anchor by the chain). Not to be confused with bridal; I used to wonder what ropes had to do with weddings.

- **Buddy boating:** the practice of traveling with another boat or boats. It provides companionship while traveling, safety in numbers, and the possibility—or illusion—of help nearby. If taken to an extreme, a group of buddy boats will divide the labor of trip planning, weather research, and provisioning—it can devolve into group think and reduces self-sufficiency and proper preparation.

- **Code zero:** Sounds like an emergency procedure but is actually a large triangular racing sail made out of high tech material, used for reaching in light wind conditions or for downwind sailing in conditions too strong for the spinnaker (in photos of boats, the spinnaker is the pretty multicolored one made out of parachute material).

- **Cruising kitty:** The money used to fund sailing trips, not to be confused with the liveaboard cat! A couple may stay put somewhere to work and fill the cruising kitty before setting sail again.

- **Dragging anchor:** What happens when the anchor doesn't set (or reset when the wind or tide changes). If it isn't dug into the bottom, the boat will move with the wind or tide, dragging the anchor across the bottom and potentially resulting in running aground (see Running Aground).

- **Galley:** Boat kitchen. Thus, the cook might be considered the galley slave. Thankfully, no rowing required.

- **Genoa:** Normally, a type of salami, but on a boat, a large triangular sail attached to a stay that runs from the bow of the boat to the top of the mast. Larger than a jib but serves the same function. Also called *genny* or, if very large, *gennaker* (cross between genoa and spinnaker, not unlike a code zero).

- **Ground tackle:** Anchor and chain. Anchor can be called the hook. Alternately, what you feel like doing to the captain or crew when they do not communicate clearly while anchoring.

- **Head:** Boat toilet. You might say, "I need to use my head." Or, "My head is not working properly." Or, "Your head is smelling a little funky, maybe you should clean it." Or, with a large family on a boat, "Two heads are better than one!"

- **HEAVING TO:** A magical position into which you maneuver your boat during a storm so that the wind in the sails counters the direction of the rudder and everything holds still. Like pushing pause. Not to be confused with dry heaving, which happens after you have been seasick for the better part of a day and can't even hold down water.

- **HEELING OVER:** What monohulls do when the wind exerts force on the sail; tipping. This is why all the lockers and cupboards on a monohull are locked and everything must be stowed when sailing. On long passages, one is required to live on an incline; one reason why we bought a catamaran.

- **HELM:** The steering wheel or tiller. If you're at the helm, you're in charge of steering, and you would then be the helmsman. Could also be a helmswoman or helmschild, but that doesn't sound as nautical.

- **IN THE DRINK:** Going for an unintended swim. One of my favorite T-shirts had printed on it, "Better to be on a boat with a drink on the rocks than to be in the drink with a boat on the rocks."

- **In the lee/on the lee:** In the lee means you have finally rounded the corner and anchor for the night on the leeward side of the island, the side away from the wind relative to the island, where a boat is somewhat protected. On the lee is when the wind shifts overnight and you wake up in a panic because the boat is slowly dragging toward the lee shore, relative to the boat.

- **Jibe:** When the stern (back) of the boat turns through the wind. Like a downwind tack. Sometimes happens accidentally and can be loud and dangerous (see "boom"). When running downwind, one might rig a preventer, a line that keeps the boom from being able to swing and thus prevents a jibe.

- **Lazarette:** a locker or storage area on a boat, usually below decks in the aft section.

- **Leeward/windward:** the leeward side of an object is that which faces away from, or is protected from, the wind. The windward side is self-explanatory. Also, Leeward and Windward Islands in the Caribbean, which run from the Virgin Islands to Trinidad.

- **Lifelines:** a barrier of thin cables that runs down the sides of the boat to create an illusion of safety. They are attached by stanchions and are under tension, so that one can use them as a railing if needed. More often used as laundry lines.

- **Man overboard:** When the man on watch in the

middle of the night decides to pee overboard and doesn't hold onto the boat tightly enough. The boat may sail on for hours without anyone noticing. The first rule of falling overboard is never fall overboard.

- **NAKED WHITE MAN DANCE:** When it starts to rain in the middle of the night and couples all over the harbor come out on deck in the buff, running around battening the hatches.

- **ON A PLANE:** when a boat experiences hydrodynamic lift, zooming along on top of the water instead of plowing through it; like flying, but not to be confused with riding on an aircraft.

- **ON THE HARD:** a boat on land, usually on blocks (of wood) or jack stands (metal supports) so that the bottom can be worked on. Also, *hauled out*. Not to be confused with *on the rocks* or *hard aground*, which means that boat has gone where it ought not to have gone.

- **PORT:** 1. The place to or from which one sails. 2. The left-hand side of the boat when you are facing the bow (front). Sailing instructors have all sorts of helpful mnemonic devices for helping one remember: port and left both have four letters. Also, Port wine is red and you keep red markers/lights on the right when returning to a port (red-right-returning). You pass other boats port to port, staying on the right side of the road. At night, that means you will see the other boat's red light. If you see green, something is wrong. See starboard.

- **Provisioning:** Shopping for provisions, food, and supplies for a sailing trip; going to the grocery store. Usually involves loading groceries multiple times: into the cart, onto the counter at checkout, into the waterproof bags, into the trunk of a car or hand wagon, into the dinghy, onto the deck of the sailboat, into the cockpit, into the galley, and into various lockers and hidey holes. Takes all day, breaks a few eggs, and requires a cold beverage to recover, especially in the tropics.

- **Reefing:** Reducing sail area (making the triangle smaller) by lowering the sail. Supposed to be done before the strong wind shows up, but often done after, and in a bit of a panic.

- **Refit:** To work on or complete a major upgrade on a boat. Alternately, to measure something incorrectly, like the new refrigerator, which requires trimming cabinetry to install it: "We were in the middle of a major galley refit when we discovered the fridge didn't fit, so we had to refit it."

- **Running aground:** taking the boat too close to shore or to a shoal and hitting the bottom. *Hard aground* is when you are actually stuck in the sand or on the reef and can't get off without waiting for high tide or help. Every sailor has run aground. If they say they haven't, it's because they have either never left the dock or because they are lying.

- **RHUMB LINE:** Course between point A and point B, a straight line drawn on the chart. Also, *rum* line—the most efficient path to your destination and celebratory drink at anchor. Does not account for wind, current, or depth.

- **SALON:** The boat living room. Not to be confused with the place you go to get your hair or nails done. Also called *saloon*, not to be confused with a bar in the Old West. The sofa in the salon is called a settee.

- **SHEETS:** Lines, or ropes, which are used to adjust the sails of a boat. Also, oddly shaped pieces of fabric which are supposed to cover the bunks, or beds, on a boat, but which are inevitably the wrong size or shape, so that making the beds becomes a wrestling match done in a small, hot space.

- **SPINNAKER:** a large, lightweight sail that balloons out in front of the boat on a downwind sail (wind coming from behind). These are the beautiful, colorful sails you see in photographs of sailboat races. Alternately called the kite or 'chute for its resemblance to a parachute. Also notorious for being difficult to deploy, collapsing in a wind shift, twisting itself up, falling in the water when doused (or brought in), and generally wreaking havoc.

- **STARBOARD:** 1. The right-hand side of the boat when

you are facing the bow (forward). The light on the starboard side is green. Green markers, however, will be on your left as you return to port. If you get this wrong, you will end up in some kind of trouble. 2. A fabulous plastic material you can use to repair things on the boat, comes in sheets like plywood.

- **S/V:** Short for Sailing Vessel; a sailboat. M/V is short for Motor Vessel; a powerboat, mega yacht, trawler, etc.

- **Tack:** When the bow (front) of the boat turns through the wind. Best done when you have some speed. The sheets are eased, the boat is turned, the wind fills the other side of the sails, then sheets are tightened. Tacking is zigzagging upwind.

- **Trampolines:** An area on the foredeck of a catamaran that provides lounging area, like large, taut nets covering the space between the hulls, not to be mistaken for the springy things in the backyard of a house.

- **Weather window:** weather conditions favorable to getting to a certain destination; the right wind speed and direction for the duration of a projected passage. Not to be confused with something that protects you against weather, but rather, something that can close unexpectedly and leave you stuck right where you are until the next window opens.

- **Yacht club:** A place generally to be avoided if you

are sailing your floating house full of children around. Usually has a nice bar and restaurant, often attached to a nice marina, and sometimes full of wealthy people who rarely untie the lines that keep the yachts tied to the dock. In Puerto Rico, pronounced "yatch cloob."

# ACKNOWLEDGMENTS

The following people made this book possible:

My nonconforming parents and siblings. You told me I could do whatever I set my heart on and I believed you. I am grateful for the weirdness of my childhood.

My fourth grade teacher at Purple Sage Elementary School, Anderson Mill, Texas, 1984. Miss Davis, wherever you are, thank you for encouraging me to write.

My best friend from Middlebury College, Heather Sanborn, who gave me my first real sailing lesson in a Sunfish on a lake in Maine.

Captain Josie Longo, who taught me basic sailing and gave me the courage to write about sailing.

Kimberly Ward of *S/V Ally Cat*, who wrote about her own sailing adventures and inspired me to write mine.

Summer Herd, friend and former liveaboard, whose accountability and partnership were invaluable.

Andrea Marcovici, who read my rough draft and made many helpful suggestions.

John and Boni Wagner-Stafford at Ingenium Books, whose

encouragement and advice helped elevate this manuscript into a book.

My tribe of supportive family and friends—sailors and landlubbers alike. Especially my understanding husband, Jay, and five children, Eli, Aaron, Sarah, Sam, and Rachel, without whom this story would not exist.

Crew of s/v Take Two, Rio Dulce, 2018

# ABOUT THE AUTHOR

Tanya Hackney graduated with a B.A. from Middlebury College in 1997, with a major in English and double minor in French and Education. She taught kindergarten in Atlanta, Georgia, then she homeschooled her five children while living full time aboard the sailboat, *Take Two*. She learned to sail in 2007 and did the coursework for ASA101 and ASA103 after attending a women's sailing seminar in St. Petersburg, Florida. She's lived aboard, traveled, and written for the sailing blog www.taketwosailing.com for more than a decade. Tanya has always had a bad case of wanderlust, taking countless road trips as a child, spending a semester abroad during college, and honeymooning in Central America. In her free time, she plays the ukulele, paints landscapes, and kayaks. She wrote her first story at age six, but *Leaving the Safe Harbor* is her first published full-length work.

Printed in the USA
CPSIA information can be obtained
at www.ICGtesting.com
JSHW022111130923
48138JS00005B/141